PRINCIPLES OF
STRUCTURE

5th EDITION

PRINCIPLES OF
STRUCTURE
5th EDITION

KEN WYATT • RICHARD HOUGH

CRC Press
Taylor & Francis Group
Boca Raton   London   New York

CRC Press is an imprint of the
Taylor & Francis Group, an **informa** business

CRC Press
Taylor & Francis Group
6000 Broken Sound Parkway NW, Suite 300
Boca Raton, FL 33487-2742

© 2013 by Ken Wyatt and Richard Hough
CRC Press is an imprint of Taylor & Francis Group, an Informa business

No claim to original U.S. Government works

Printed in the United States of America on acid-free paper
Version Date: 20121107

International Standard Book Number: 978-0-415-66727-2 (Paperback)

Library of Congress Cataloging-in-Publication Data

Wyatt, K. J. (Kenneth James), 1935-
    Principles of structure / authors, Ken Wyatt, Richard Hough. -- Fifth edition.
       pages cm
    Includes bibliographical references and index.
    ISBN 978-0-415-66727-2 (hardcover : alk. paper)
    1. Structural analysis (Engineering) I. Hough, R. (Richard) II. Title.

TA645.W92 2013
624.1'7--dc23   2012035177

Visit the Taylor & Francis Web site at
http://www.taylorandfrancis.com

and the CRC Press Web site at
http://www.crcpress.com

Contents

Introduction

Welcome to the 5th edition of *Principles of Structure*. This edition retains the style, format and most of the content of previous editions, and brings the text and examples into alignment with international practice and the mks (metre, kilogram, second) International System of units (SI). It also features six new international buildings as 'applications' of the principles described in the text.

While the computerisation and automation of structural engineering design continues apace, simple manual calculation still underpins most early stage structural design. To check the feasibility of an initial idea and to estimate approximate sizes for structural members, the manual methods explained in *Principles of Structure* remain quickest and simplest.

The text is intended primarily for undergraduate students of architecture — also students of building, and students in some project management and construction management courses. Since its first publication in 1974, it has been successful in meeting a consistent demand in these areas. While there are many introductory texts that remain entirely qualitative in their treatment of structural principles, students and teachers alike seem to prefer some quantitative coverage. Perhaps it is not possible to gain a durable appreciation of the engineer's potential contribution to design through a wholly qualitative approach. Certainly the most effective dialogue between architect and engineer occurs when the architect comprehends something of the engineer's quantitative language, and we believe this is best conveyed through simple mathematics.

The practice of structural engineering has expanded in recent years to address a range of issues beyond the core activities of structural analysis and design. These issues include: sustainability; analysis of complex geometries linked to post-modern architectural styles; and building information modeling together with the opportunities it brings for optimisation of design and fabrication.

These issues were often significant during structural design of the new building 'applications' described in this 5th edition. For example, establishing geometry, then carrying out structural analysis, joint modeling and fabrication scheduling for the steel frame of the CCTV building Beijing, and the main stadium for the 2008 Olympics, all relied heavily on computer methods, with various degrees of interoperability between architectural, engineering, and fabrication models.

Sustainability goals pose a multitude of challenges for designers of structures. Structural engineers are nowadays invited to join multi-discipline design discussions about the environmental implications of structural materials, such as recyclability, and thermal considerations leading on to building energy efficiency. For example, the Serpentine Pavilion, the Raleigh-Durham air terminal building, and Druk White School, Ladakh, all took advantage of the lower global warming potential of timber as a structural material, and all offer ease of disassembly for potential recycling at end of life.

More information on the building applications in this edition can be found on-line, often at the website of the project architect, or of design firm Arup, who were structural engineers for all of the applications.

All these additional aspects of holistic building design serve to enrich the contemporary practice of structural engineering. However, it remains true that good structural design still relies first and foremost on a clear understanding of the structural principles outlined in *Principles of Structure*. Understanding the effect of applied loads, and providing an efficient arrangement of structural material to respond to these effects, is still the fundamental process of structural design.

Two particular features of the book need mentioning. Firstly, the subject matter is presented in two parallel streams: a text and a commentary, as described in the section on How to Use This Book. Secondly, Section 1, dealing with forces and their effects, forms the basis on which the other sections are built, so needs to be thoroughly understood at the outset.

We hope you will find that the format and style of *Principles of Structure 5th edition* make it easy to use, that it leads to a clear understanding of basic structural behaviour, and that it stimulates your interest to pursue further the absorbing study of structural design.

Ken Wyatt
Richard Hough

Authors

Ken Wyatt was the mainstay of structural engineering teaching in the Faculty of the Built Environment at the University of NSW for many decades, where he developed the original text of *Principles of Structure*. He is also a structural engineer, researcher, materials scientist and heritage consultant.

Richard Hough has practised structural engineering in London, Los Angeles and Sydney, working for Arup with many of the world's best-known architects on many of the world's most notable buildings. He is presently a Principal in the firm's Sydney office.

How to Use This Book

The subject matter of this book is divided into two parts which are presented in parallel format — the text and the commentary.

The *text* is on the right-hand pages, and seeks to develop each topic in the logical, rigorous way that you would expect to find in any standard text. On the left-hand pages you will find a *commentary* to the text. In the commentary you will find supplementary explanations and expansions of points made in the text. There are also hints and suggestions, problem sheets and worked examples. The commentary is not sequential; it does not tell a story or develop a theme.

This book is designed to be used in a particular way. When you approach a new chapter, just read through the text on the right-hand pages. Perhaps there may be parts you don't fully understand, but don't let that stop you; skip over them and keep reading to the end of the chapter. You will then have a pretty good idea of what the topic is about, even though some bits may be a little hazy.

After a day or two, start to read the text again, more slowly this time, making sure that you understand all of it. At the same time, read and understand the commentary, item by item. Keep cross-referencing from text to commentary; they look at the same issues from different viewpoints. Study each worked example carefully, so that you can see the reasons for each step.

For your third reading, ignore the text and, instead, study the commentary. Treat each of the worked examples as a set problem, and work it through on a separate page. Don't refer to the worked solution until you reach a stage where you are unable to progress further. Then go back to the text to clear up the point that blocked you.

By the time you've mastered all the worked examples in the section in this way, you are ready to test yourself against the worksheet at the end of the chapter. Don't start on the worksheet until you are able to do all of the worked examples! Check your answers against those starting on page 187.

The Appendix contains a separate section called 'A First Encounter with Statics'. This is a supplement to Chapter 1, intended for those students unfamiliar with the basic concepts of forces and their effects. Use this supplement if you need a fuller explanation of the basics.

There are many texts on structures or structural mechanics and each has its own particular strengths. If one topic baffles you, there will be another book that has the special key for this topic for you. Do not hesitate to supplement *Principles of Structure* whenever you need additional information or fresh insight.

Principles of Structure is of course intended as an introductory text. The methods explained are rudimentary and the data provided is approximate. For the structural design of real structures, the services of a suitably qualified structural engineer should be obtained. The authors accept no responsibility for consequences arising from the application of information in the text.

1 Forces, Moments and Equilibrium

NEWTON'S LAWS

Newtonian physics is based upon the following three laws:

1. A body at rest will remain at rest and a body in motion will remain in motion with constant velocity unless an external resultant force acts upon the body.

2. The rate of change of momentum of a body is proportional to the external resultant force acting on the body. If the mass of the body remains unchanged, the product of the mass and the acceleration of the body equals the external force.

3. For every external force that acts on a body (the action force) there is an equal and opposite force (the reaction force) that acts on some other body.

It is worth noting that Newton's Laws refer to *external* forces. If you place your pen on the table in front of you, there are two external forces acting on it. One is its weight, the gravitational force caused by its mass. If the pen has a mass of 0.02 kg, what is its weight? What is the direction of this gravity force, and what is its sense? The second external force acting on the pen is the thrust exerted *by* the table. What is the magnitude, direction and sense of this force? Now, there are many other forces associated with the pen, but these other forces are all *internal* forces. The clip presses against the cap, and the cap exerts a reaction force against the clip. The nib or ball-point is clamped into its mounting. Even the molecules are exerting forces upon one another. But, as long as it is the pen as a whole that we are considering, these forces are internal forces, and must not be lumped together with the two external forces if we are to successfully apply Newton's laws to deduce anything about the external forces.

CONTENT OF CHAPTER 1

Chapter 1 deals very briefly with the external forces that act on rigid bodies in equilibrium, and with the effects produced within those bodies. You will have studied some or all of this material before — that is the reason for the brevity of the treatment in this chapter. *However, if you do encounter any difficulties, refer to the Appendix.* Most of the material will be amplified later in the text, but because this chapter is so very fundamental to your future studies, you must be thoroughly conversant with it. *Make quite sure that you clearly understand all the material in Chapter 1 before you proceed further with your studies.* Read other texts to clear up any points of doubt; often a different viewpoint will provide a new insight.

1.1   FORCES ON STRUCTURES

In considering the structure of buildings, we are concerned with the effects of the forces that will be applied. These forces include the pressure of wind, the weight of rainwater on the roof, the weight of the occupants, the forces produced by earthquakes, even the weight of the building itself. Our task as designers or builders is to ensure that the building is able to withstand these forces without deforming excessively or collapsing.

1.2   MEASUREMENT OF FORCE

Forces are measured in Newtons (N). A force of one Newton applied to a mass of one kilogram (kg) will produce an acceleration of one metre per second squared (m/s²). The forces exerted on buildings are fairly large, and we will adopt the kilonewton (kN = 10^3 N) as our basic unit of force. The acceleration produced by the gravitational attraction of the earth is 9.81 m/s²; hence the gravitational force (weight) acting on a mass of M kg is 9.81 M Newtons (Figure 1.1). Approximately then, a mass of M kilograms supported by a structure will apply a force of 10 M Newtons to that structure.

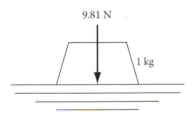

FIGURE 1.1

1.3   EXTERNAL FORCES

To fully describe a force, however, we need to specify more than just its magnitude. The following characteristics are required:

FORCES AND THEIR EFFECTS

The study of forces acting upon objects at rest is called 'statics'; a knowledge of the principles of statics is essential for an understanding of structure. Often, though, it is easier to understand these principles if one tries to envisage the *motion* that might be produced by a system of forces. Once you can imagine the way an object moves or deforms, you are well on the way to being able to evaluate the forces needed to maintain equilibrium of the object.

The following terms have been defined in the text:

— resultant, component, moment, couple, equilibrium

Use as many of these terms as possible to explain what happens in each of the situations sketched below.

Using a broom on a smooth floor

Pushing a wheelbarrow over a low step

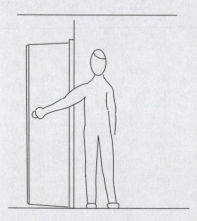

Turning a door-handle and opening the door

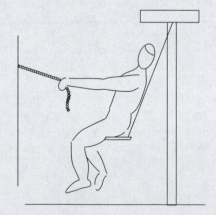

Tugging a rope while sitting on a swing

(a) *the magnitude*;
(b) *the direction* with respect to some fixed frame of reference (e.g. vertical, horizontal, north-west, etc.) and the *sense* in the specified direction (e.g. for a vertical force, the sense would be upwards or downwards; for a force in a north-west direction, the sense would depend upon whether the force was directed *to* the north-west, or *from* the north-west). When we represent forces in diagrams we have a convention that forces directed either upwards or to the right have a positive sense while those acting downwards or to the left are negative;
(c) *the location*, usually specified by the co-ordinates of a point through which the force passes.

1.4   RESULTANTS AND COMPONENTS

On occasion, we may wish to replace two or more forces by a single force that produces the same effect, i.e. which is equivalent to the sum of the other forces. The single equivalent force is called the *resultant* of the other forces. In Figure 1.2, F_1 produces the same effect as F_2 and F_3 together; F_1 is the resultant of F_2 and F_3.

If we represent each force by a vector (an arrow having the same direction and sense as the force, and whose length represents the magnitude of the force to some scale) we can find the resultant of two forces by a graphical construction known as the parallelogram of forces (Figure 1.3).

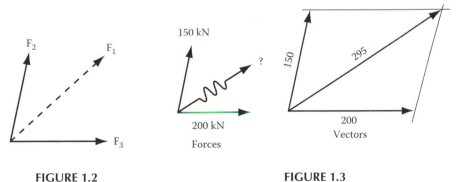

FIGURE 1.2   **FIGURE 1.3**

In the reverse of this process, we sometimes wish to replace a single force by two other forces; the other forces are then called the *components* of the single force. In Figure 1.2, F_2 and F_3 are components of F_1, i.e. the single applied force F_1 could be replaced by the pair of forces F_2 and F_3, and the effect on the body would be unchanged.

By using the parallelogram of forces, we can see that any one force can be replaced by an infinite number of pairs of component forces. F_1 and F_2, or F_3 and F_4, or F_5 and F_6, are all component pairs of the force F (Figure 1.4).

EXAMPLE 1.1   CONCURRENT FORCES

A crane-hook H is held in equilibrium by two masses and by a fixed cable HA. By considering the equilibrium of the hook, calculate the force in HA. The gravitational forces acting on the masses of 163 kg and 200 kg are 1630 N and 2000 N respectively.

Consider the equilibrium of the hook H. The three forces acting on the hook are as shown, force HA being unknown in both magnitude and direction.

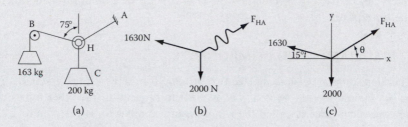

For equilibrium, $\Sigma F_x = 0$

i.e. $\Sigma F_x = -1630 \cos 15° + F_{HA} \cos \theta = 0$

i.e. $F_{HA} \cos \theta = 1578$ N   (1)

Also, $\Sigma F_y = 0$

i.e. $\Sigma F_y = +1630 \cos 75° + F_{HA} \cos (90 - \theta) - 2000 = 0$

i.e. $F_{HA} \sin \theta = 2000 - 422 = 1578$ N   (2)

Combining equations (1) and (2), we have:

$$\frac{\sin \theta}{\cos \theta} = \frac{1578}{1578} = \tan \theta$$

$$\therefore \theta = 45°$$

$$\therefore F_{HA} = \frac{1578}{\cos 45°} = +2230 \text{ N}$$

The unknown force in HA acts at 45° to the horizontal, upwards to the right. Its magnitude is 2230 N.

The reader should note particularly the way in which the sign convention for sense (stated in para. 1.3) has been used in this example. Frequently, the sense of an unknown force is not obvious at the start of a problem. *In these cases, a sense should be assumed; if the numerical solution is positive the assumed sense was correct; if negative, the assumption was incorrect and the force has the opposite sense to that assumed.*

The vector notation used to represent the force F_{HA} is also worth noting. This 'wriggly arrow' is a reminder that the direction of this force is not yet known; the direction must be found as part of the solution.

Often we are concerned with two perpendicular components of a given force, e.g. what are the vertical and horizontal components of an inclined force? If force F_1 is to be replaced by two perpendicular components F_2 and F_3 acting in directions 'a' and 'b' at point A (Figure 1.5), we see that the parallelogram of forces becomes a rectangle and that $F_2 = F_1 \cos \theta$ and $F_3 = F_1 \cos \alpha$.

We have *resolved* F_1 into a pair of *rectangular components*. When dealing with rectangular components, *the component in direction AB of a force F in direction AC is F cos CAB* (Figure 1.5(c)).

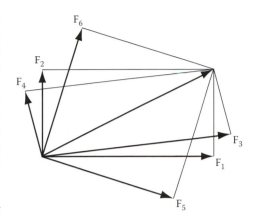

FIGURE 1.4

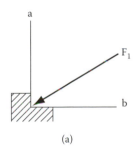 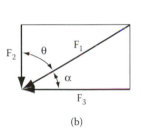 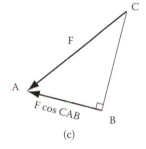

(a)   (b)   (c)

FIGURE 1.5

1.5   MOMENTS

When an external force acts upon a body it may tend to cause a translation (displacement in a straight line) or a rotation. In most instances both tendencies are present, and we need a concept to measure the *turning effect* or rotating effect of a force about a point. The *torque* or *moment* of a force about a point is the product of the force and the *perpendicular distance* from the point to the force (Figure 1.6). The perpendicular distance is called the 'lever arm' of the force. Being the product of force and length, moments have the units of Newton-metres, Nm, or kNm, etc. It is most important to note that we may measure the moment of a force about *any point we choose*; the point does not have to be the centre of gravity, the fulcrum, the point of support or any such particular point. The moment of a force about a point is simply a measure of the turning effect produced by *that* force about *that* point. We shall adopt a convention that *clockwise moments are positive*, and anti-clockwise moments are negative.

WORKSHEET 1.1

1.1 A force of 10 kN is applied at an angle of 30° to the horizontal (Figure 1).

    (a) *Calculate* the component of this force in each of the directions AB, CD and EF.

    (b) Confirm for yourself, by rapid sketches drawn approximately to scale, that the values you have computed seem reasonable.

1.2 Figure 2 shows three concurrent forces acting on a body.

    (a) Are these forces in equilibrium?

    (b) If not, what additional force would need to be applied to the body in order to produce equilibrium?

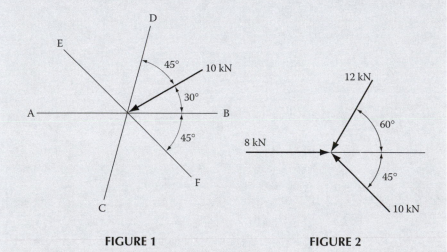

FIGURE 1   FIGURE 2

1.3 Figure 3 shows an awning erected at the front of a suburban shop. If we assume that the weight of the awning is concentrated at its mid-point:

    (a) What is the moment of this weight about the hinged support at A?

    (b) The force in the steel bar BC will also produce a moment about A. What is the lever arm of this moment?

    (c) What force does the steel bar BC exert on the awning?

    (d) What force does the hinge A exert on the awning? What are the vertical and horizontal components of this force?

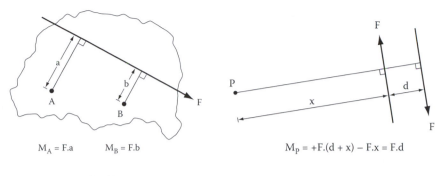

$M_A = F.a$ $M_B = F.b$ $M_p = +F.(d + x) - F.x = F.d$

FIGURE 1.6   **FIGURE 1.7**

1.6   COUPLES

A couple consists of two equal, parallel, non-collinear forces of opposite sense. If we measure the total moment produced by the two forces about any point in the plane of the couple, we find that the moment of a couple about any point in its plane is constant, and is equal to the product of one of the forces and the perpendicular distance between the forces (Figure 1.7). The perpendicular distance is called the 'arm' of the couple.

1.7   EQUILIBRIUM

Consider a rigid body being acted upon by a system of external forces (Figure 1.8). If the body does not accelerate horizontally, there can be no horizontal resultant force acting on the body (i.e. the sum of the horizontal components of all forces in the system must be zero).

Similarly, if the body does not accelerate vertically, there is no vertical resultant force. If the body does not rotate about any point, then the resultant moment of all forces about any point must be zero. Consequently, if the body has no components of acceleration in either vertical or horizontal directions, and no rotation about any point, it is said to be *in equilibrium* under the action of the forces acting upon it.

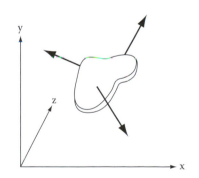

FIGURE 1.8

Thus, having stipulated three mutually perpendicular axes, x, y and z, the necessary and sufficient conditions for a body to remain in equilibrium are that:

(a) the sum of force components in the x direction is zero. $\Sigma F_x = 0$
(b) the sum of force components in the y direction is zero. $\Sigma F_y = 0$

WORKSHEET 1.1   (CONTINUED)

1.4 Harry the painter (who weighs 80 kg) is standing halfway up an aluminium ladder of negligible weight (Figure 4). Assume that the thrust applied by the wall to the ladder is horizontal.

   (a) What is the vertical component of the force exerted on the ladder by the floor? (Consider equilibrium of vertical forces.)
   (b) What is the thrust exerted by the wall on the ladder? (Consider equilibrium of moments.)
   (c) What is the resultant force exerted by the floor on the ladder?

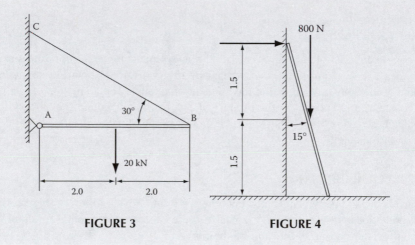

FIGURE 3   **FIGURE 4**

(c) the sum of force components in the z direction is zero. $\Sigma F_z = 0$
(d) the resultant moment about *any* point is zero.

For the present, we will be considering co-planar force systems acting on two-dimensional structures. For a structure of this type to be in equilibrium, the following conditions must be satisfied:

$$\Sigma F_x = 0 \quad \Sigma F_y = 0 \quad \Sigma M = 0$$

In applying these equations to a system of forces, one must be quite certain that the forces being considered *all act upon a single body*. It is a common mistake to include, within the one system, both forces acting *on* the body and also forces exerted *by* the body on some other object or component. *The first step is to identify the body being investigated, and then consider only those forces which act upon the body.*

1.8 CO-PLANAR FORCES

Any system of co-planar forces acting on a rigid body in equilibrium must satisfy these three equations of equilibrium. Conversely, if a body is in equilibrium under a system of forces, some of which are unknown, we may use these three equations to determine the unknown forces by solving the equations simultaneously. Note especially that there are only *three equations*; therefore we can solve for only three unknown force characteristics. These three unknowns, for example, could be any of the following (Figure 1.9):

(a) the *magnitude, direction* and *location* of a single force;
(b) the *magnitude* and *direction* of a force (for which we already know the location), and the *magnitude* of a force (for which we know the direction and location);
(c) the *magnitudes of three forces*, for which all other information is known.

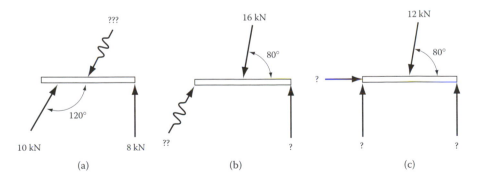

FIGURE 1.9

EXAMPLE 1.2   NON-CONCURRENT FORCES

A canopy for an outdoor stage consists of a series of bent girders, each supported by a steel cable. When each girder is carrying the loads shown, calculate the force in the cable and the reactions at the hinge.

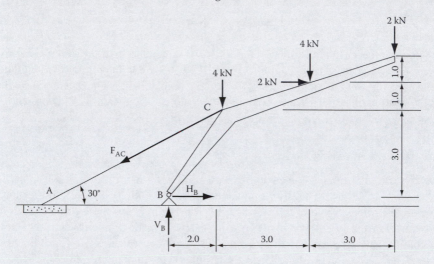

The hinge at B can supply two reaction forces: a vertical component and a horizontal component. The total number of unknown forces is therefore three (two at B and one at A); they can be determined by the equations of equilibrium and the reactions are statically determinate.

By application of the equations of equilibrium, we have:

$$\Sigma F_y = 0 \qquad -F_{AC} \cos 60 + V_B - 4 - 4 - 2 = 0$$
$$\text{i.e. } V_B - 0.5 \, F_{AC} = 10 \tag{1}$$

$$\Sigma F_x = 0 \qquad -F_{AC} \cos 30 + H_B + 2 = 0$$
$$\text{i.e. } H_B - 0.866 \, F_{AC} = -2 \tag{2}$$

$$\Sigma M_C = 0 \qquad +V_B \times 2 - H_B \times 3 + 2 \times 1 + 4 \times 3 + 2 \times 6 = 0$$
$$\text{i.e. } 2V_B - 3 \, H_B = -26 \tag{3}$$

These three equations may now be solved simultaneously by the usual techniques; e.g. equations (1) and (2) may be combined to yield:

$$1.732 \, V_B - H_B = 19.32$$

which, when combined with equation (3) gives the result:

$$V_B = 26.2 \text{ kN}$$
$$H_B = 26.1 \text{ kN}$$
$$F_{AC} = 32.4 \text{ kN}$$

1.9 CONCURRENT FORCE SYSTEMS

In Example 1.1, the force system was a *concurrent co-planar* system, i.e. the forces all passed through the same point and all acted in the same plane. Force systems are quite frequently encountered which are *non-concurrent* (see Example 1.2), in which not all the forces pass through a single point.

If a body is in equilibrium under the action of only three forces, the forces are either concurrent or parallel. This principle, sometimes called the 'Three-Force Condition', is very important in structural analysis, and has a wide application. How would you prove it yourself? (Hint: Assume that two of the forces are either concurrent or parallel, and then deduce the characteristics of the third force in each instance by application of the equations of equilibrium.)

1.10 FORCES AT SUPPORTS

The forces that are applied to a structure by actions such as those discussed in paragraph 1.1 are called *loads*, and will be discussed later.

Of course, for the structure to remain in equilibrium, the supports must also apply forces. These supporting forces are called *reactions*, and they must be computed before the structure itself can be analysed.

In planar structures we are concerned with three types of support:

(a) a *hinge* or *pin* is capable of providing *two reaction components* — it can prevent motion in two perpendicular directions, but permits rotation about its axis;
(b) a *roller*, used in conjunction with a hinge, can provide only one reaction component — it can prevent motion in a direction perpendicular to the surface on which the roller rests, but permits motion parallel to this surface and also permits rotation;
(c) a *fixed support* can provide *three reaction components* — it prevents motion in two perpendicular directions and prevents rotation.

Thus, a roller provides a single reaction component (a force of known direction and location but unknown magnitude); a pin supplies two components (two forces of known direction and location but unknown magnitude OR a single force of known location but unknown magnitude and direction); and a fixed support supplies three components (two forces and a moment).

The conventional method of representing these supports is shown in Figure 1.10. It should be noted that the reaction force provided by a roller may have either of two possible senses. In Figure 1.10(a), the reaction may be either upwards or downwards, and the roller itself is considered to be capable of transmitting either compression or tension.

THE USE OF THE EQUATION FOR EQUILIBRIUM OF MOMENTS

In Example 1.2, the same result could have been obtained by applying the equations in other ways. The equation for moments, in particular, is a very powerful tool, and the selection of the point about which moments are computed is most important. For example, if we had taken moments about B, the force F_{AC} could have been obtained directly, because the other two unknowns pass through B and would not appear in the resulting equation. Similarly, if moments had been taken about the point of intersection of the vertical line through B and the cable AC, only H_B would have appeared in the equation. Finally, V_B could be obtained by taking moments about A. We see that, although the equations of equilibrium will always result in a solution for a determinate structure, the use of foresight in their application will often reduce the labour required.

EXAMPLE 1.3   REACTION FORCES FOR A BEAM

A beam ABC is supported by a hinge at A and a roller at B and carries the loads shown. Determine the reactions.

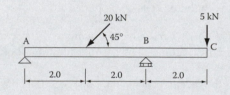

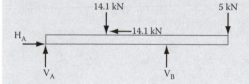

Because there are only three unknown forces, the system is determinate.

$$\Sigma F_y = 0$$
$$+ V_A + V_B - 14.1 - 5 = 0$$
$$V_A + V_B = 19.1 \qquad (1)$$

$$\Sigma F_X = 0$$
$$+ H_A - 14.1 = 0$$
$$H_A = 14.1\,kN \qquad (2)$$

$\Sigma M_A = 0$, i.e. $14.1 \times 2 + 5 \times 6 - V_B \times 4 = 0$ $V_B = \dfrac{58.2}{4} = 14.6\ kN$

$\Sigma M_B = 0$, i.e. $V_A \times 4 - 14.1 \times 2 + 5 \times 2 = 0$ $V_A = \dfrac{18.2}{4} = 4.6\ kN$ $\qquad (3)$

Substitution of V_B and V_A into equation (1) provides a check that no arithmetical error has occurred.

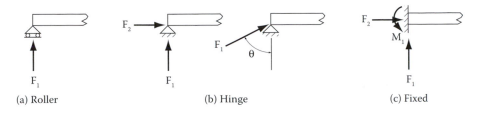

(a) Roller   (b) Hinge   (c) Fixed

FIGURE 1.10

1.11 STABILITY AND DETERMINACY

Because there are *three* equations of equilibrium, a body must be supported by at least *three* reaction components if it is to remain stable. A body supported by less than three reaction components is *statically unstable* under a general system of co-planar forces (Figure 1.11).

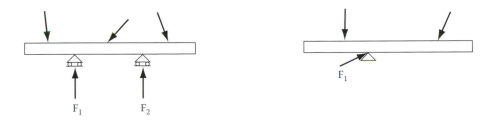

FIGURE 1.11

If a two-dimensional structure has exactly three unknown reaction components, the reactions may be obtained by simultaneous solution of the three equations of equilibrium and are said to be *statically determinate*. If there are more than three reaction components, they cannot be determined from equilibrium alone and are *statically indeterminate* (Figure 1.12).

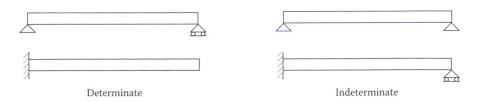

Determinate   Indeterminate

FIGURE 1.12

EXAMPLE 1.3   (CONTINUED)

Example 1.3 illustrates a standard procedure that should be adopted for beams. The two vertical reactions are computed independently by taking moments about the points of support, and the values obtained are checked by considering the sum of the vertical forces.

INSTABILITY

In paragraph 1.11 we saw that a structure supported by less than three reaction components will be unstable. Occasionally, structures are encountered that do have three reaction components, but these are arranged geometrically in such a way that the structure is still unstable. Such structures are said to be geometrically unstable.

EXAMPLE 1.4

Determine the resultant of the system of four parallel forces shown in the diagram.

Magnitude of resultant:

$= 10 + 20 + 5 + 5 = 40$ N

This resultant must be located in a position such that it will produce a moment about any point equal to the sum of the moments produced by the parallel force system.

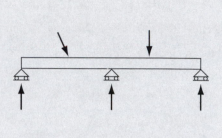

For example, for moments about point A:

$40 \times d = 20 \times 40 + 5 \times 80 + 5 \times 100$

$\therefore d = \dfrac{1700}{40} = 42.5$ mm

Hence the resultant of the system is a force of 40 N, parallel to the other forces, and located 42.5 mm from A.

1.12   PARALLEL FORCE SYSTEMS

We saw in paragraph 1.4 that a system of concurrent forces can, in general, be replaced by a single force (the resultant) which produces the same effect as the system. Similarly, a system of *parallel* forces can be replaced by a resultant force. To adequately specify the resultant we need to know its magnitude, direction, sense and location.

Consider a weightless body under the action of three parallel forces, F_1, F_2 and F_3, all having similar sense (Figure 1.13). Suppose that this body is maintained in equilibrium by a reaction-force P. Then,

(1) From $\Sigma F_x = 0$:
F_1, F_2 and F_3 have no components in the x-direction and hence P has no components in the x-direction, i.e. *the reaction P is parallel to the forces* F_1, F_2 *and* F_3.
(2) From $\Sigma F_y = 0$:
$P = F_1 + F_2 + F_3$, i.e. the *reaction P has a magnitude equal to the sum of the magnitudes* of F_1, F_2 and F_3.
(3) From $\Sigma M = 0$:
The sum of all moments of all forces about any point must be zero, and the location of the reaction P can be found by taking moments about any point, e.g. the distance x can be found from the equation:

$$+ P.x - F_1.x_1 - F_2.x_2 - F_3.x_3 = 0$$

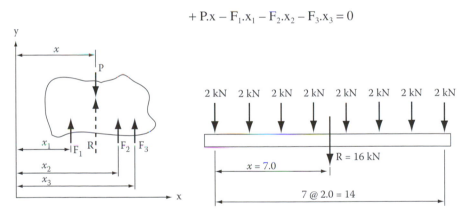

| FIGURE 1.13 | FIGURE 1.14 |

Hence, we can fully specify a *reaction* P necessary to maintain any system of parallel forces F_1, F_2, F_3, ... in equilibrium. Now the *resultant*, R, of F_1, F_2, F_3, ... will be equal and opposite to the reaction P. *Thus the resultant of a system of parallel forces is parallel to the system; its magnitude and sense are found by algebraic summation; its location is found by dividing the sum of moments about a point by the magnitude (see Example 1.4).*

EXAMPLE 1.5

The lighting standard shown has two external forces applied to it. The 100 N force is applied in the plane of the structure, and the 50 N force is perpendicular to that plane. In element AB, the 100 N force will produce bending and also shear, as can be seen by considering a plane m–m perpendicular to the axis of AB. The same force will produce bending in BC, and an axial tension of 100 N, but no shear (to the left of plane n–n there is no force parallel to n–n). DE will undergo bending and shear produced by the 50 N force, and CD has bending, shear and torsion but has no axial forces. Both external forces produce shear and bending in CF, and the 50 N force produces torsion.

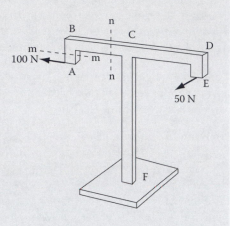

FREE-BODY DIAGRAMS

Throughout this chapter, we have been referring to forces acting on a rigid '*body*'. We have also been careful to distinguish between the forces that are *external* to the body and those that are *internal*.

Fundamental to the study of structure is the concept that we can consider *the equilibrium of just one part of a rigid structure*. We can state this concept as follows:

> If a structure as a whole is in equilibrium, each individual part must also be in equilibrium. Therefore, we can study the equilibrium of any part of a structure; and forces that are *internal within the whole structure* can be treated as *external forces upon a part* of the structure.

The best way to put this concept to use is in the Free-Body Diagram. A Free-Body Diagram is a sketch of a part of a structure showing *all* the external forces and moments acting on that part, *including* any forces or moments that the remainder of the structure exert upon the part. We have already used a Free-Body Diagram in Example 1.1, and we will study them in more detail in Chapter 4. In Worksheet 1.2, draw Free-Body Diagrams for each of the various components of the structures shown in Figures 4 and 5. Remember that the diagram *must show all forces and moments acting on the component*. The diagram for the beam BD in Figure 4, for example, will show the applied load of 1.2 kN/m, *and* the force applied by cable AD, *and* the force applied through the hinge at B.

When the force system consists of a number of equally spaced forces of equal magnitude and similar sense (e.g. as in Figure 1.14) the resultant is located midway between the extreme forces.

Quite frequently we wish to design structures to resist the effects of an applied pressure (wind pressure, water pressure, or the distributed weight of materials or people). This type of load is called a *distributed load*: the total applied force is distributed over an area (in the case of surface structures such as floors) or over a length (in the case of linear structures such as beams and girders). The most common type of distributed load is the *uniformly distributed load* U.D.L., where the *intensity* of load (N/m or N/m²) is constant.

A uniformly distributed load of 2 kN/m on a beam is represented diagrammatically as in Figure 1.15; each metre length of the beam supports a load of 2 kN. If we imagine the distributed load to be replaced by a series of equal point loads, we realise, as in Figure 1.14, that the *resultant* of this uniformly distributed load will be a single force of 16 kN applied midway along the loaded length.

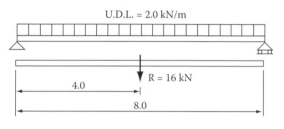

FIGURE 1.15

In general, the resultant of a distributed load is a force equal in magnitude to the total load, acting at the centre of gravity of the load. Thus, in Figure 1.16, the two distributed loads can be replaced by their resultants as two concentrated forces.

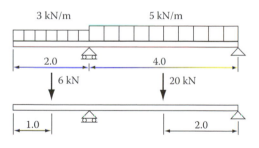

FIGURE 1.16

1.13 INTERNAL FORCES

Up to this point, we have been considering only the *external* forces acting on structures, and before we leave this introductory chapter, we should give some consideration to the types of *internal* forces that are produced inside a structure. We shall do this by examining the types of *deformations* that could be produced by the external forces.

Consider two bodies A and B connected over a common plane area x−x. We can conceive of five possible forms of deformation of B with respect to A, and these are shown in Figure 1.17.

WORKSHEET 1.2

1.1 For each of the structures shown in Figures 1, 2 and 3, calculate all reaction forces at the supports.

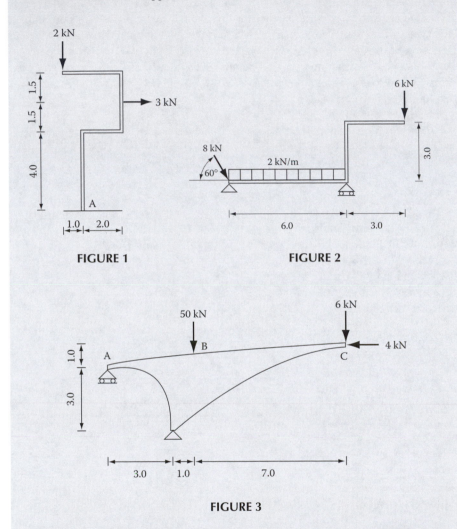

FIGURE 1   **FIGURE 2**

FIGURE 3

1.2 Figure 4 shows a cross-section of an outdoor canopy. A post ABC is rigidly restrained at the foundation at C, and supports a beam by a hinge at B and by a tension cable AD. For the uniformly distributed dead load shown:

   (a) Determine the tension in the cable AD and the force applied to the beam through the hinge at B.
   (b) Determine the reactions at C.

TYPE OF DEFORMATION	DESCRIPTION
1 Separation of A & B (stretching)	Tension
2 Increased pressure between A & B (crushing)	Compression
3 Translation (sliding) of A with respect to B (sliding)	Shear
4 Rotation of A or B about a line in the plane x–x (bending)	Bending
5 Rotation of A or B about a line perpendicular to the plane x–x (twisting)	Torsion

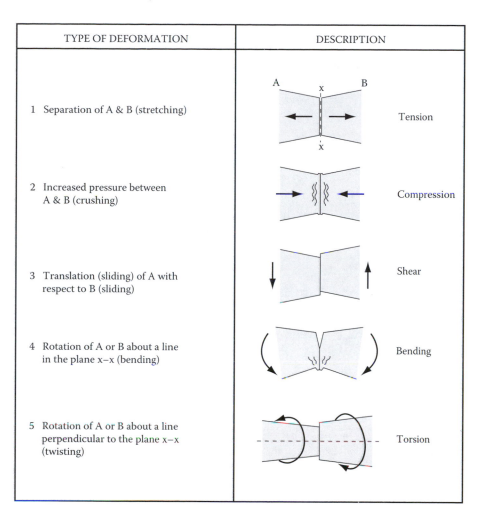

FIGURE 1.17

All possible forms of deformation of rigid bodies can be considered as combinations of these five effects: tension, compression, shear, bending and torsion.

Bending and torsion are produced by external moments, called Bending Moments and Twisting Moments. The forces that produce Bending Moments are in the same plane as the structure, and the deformation that results in the structure is curvature. Twisting Moments, on the other hand, are produced by forces lying in planes perpendicular to the plane of the structure, and the centre-line of the structure remains straight (Figure 1.18).

WORKSHEET 1.2   (CONTINUED)

1.3 A sign of mass 100 kg is erected as shown in Figure 5. Find the forces in the ropes AB and BC which are tied at B, and in the mast BD, which is hinged to a footing D. Use mathematical methods only.

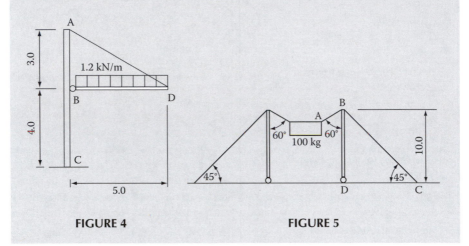

FIGURE 4   **FIGURE 5**

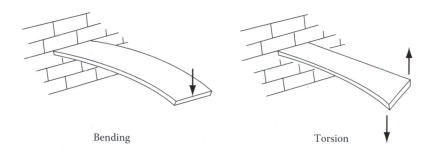

Bending   Torsion

FIGURE 1.18

What internal forces cause these deformations? The forces causing tension or compression (i.e. tensile or compressive forces) are called *normal forces* because they act normally (i.e. perpendicularly) to the surface being considered. Sometimes, these forces are also called *axial forces*, if they act in the same direction as the axis or centre-line of a part of a structure. *Internal shear forces*, on the other hand, act in a direction *parallel* to the surface being considered (Figure 1.19). The internal forces that result from bending deformations are compressive on the concave side and tensile on the convex side, i.e. pure bending results in normal forces. Torsion results in internal shear forces — forces that are parallel to the surface being considered.

The five possible forms of deformation, then, result in the formation of only two types of internal force: normal forces and shear forces. This is a most important conclusion, and our aim in all structural design is to identify and control these two simple types of internal force.

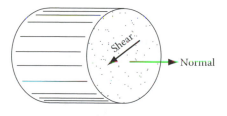

FIGURE 1.19

APPLICATION: CONCURRENT FORCES, HONG KONG AND SHANGHAI BANK (SEE PHOTOGRAPHS OVER PAGE)

There are many examples of concurrent forces in the primary structural frame of the Hong Kong and Shanghai Bank building in Hong Kong. One of the simpler ones is at the remote end of the 'outrigger trusses' that reach out beyond the main masts. The downward force from the vertical edge tension hanger is resisted by tension in the inclined hanger plus compression in the horizontal strut. Two pins were used to simplify the fabrication and construction of this remote joint. The pins are spherical in nature, to release any bending and twisting effects that might be induced by frame movements perpendicular to the plane of the outrigger truss.

Project: Hong Kong and Shanghai Bank, Hong Kong; Architect: Foster and Partners; Structural Engineer: Arup; Photos: Ian Lambot; Diagram: Arup

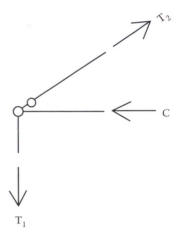

2 Ties and Struts

PRESSURE AND STRESS

We saw in the previous chapter that some external forces are applied in the form of a distributed load over a finite area of the structure. The force of wind blowing against a wall, or the weight of a heap of sand, were given as examples. When an external force is applied over an area, it exerts a *pressure* on the structure. The unit of pressure is the Pascal (Pa); one Pascal equals a force of one Newton applied over an area of one square metre. Most pressures used in structures are much larger than the Pascal, and the standard unit used is the kilopascal (kPa).

We have also seen that these external forces produce *internal* forces within the structure. These internal forces are distributed throughout the material, and the term *stress* is used for a distributed internal force in just the same way that pressure is used for distributed external forces. The unit for stress is the same as for pressure (Pa), except that in engineering materials stresses are much larger, and the megapascal (MPa) is the unit most commonly used.

$$1 \text{ MPa} = 10^3 \text{ kPa} = 10^6 \text{ Pa} = 10^6 \text{ N/m}^2 = 1 \text{ N/mm}^2$$

There are three types of stress, and they are associated with three types of deformation that may occur within a material:

(a) *Tensile stress*, which is present when the material stretches or elongates. The average tensile stress in a component is found by dividing the internal force by the cross-sectional area measured *perpendicularly* to the direction of elongation.

(b) *Compressive stress*, which is present when the material shortens or compresses. The average compressive stress is also found by dividing the internal compressive force by the cross-sectional area.

(c) *Shear stress*, which is present when the material undergoes angular distortions. The average shear stress is calculated by dividing an internal force by an area *parallel* to that force.

CONTENT OF CHAPTER 2

In order to make sure that our structures have enough strength to withstand the external forces that will be applied to them during their lifetimes, we need to be able to predict and measure the *effects* of those forces. In this chapter, we study the stress and the strain that occur in a structure when external forces are applied, and see how structural safety can be achieved by limiting these stresses. We will apply these techniques to the design of simple tension and compression structures.

2.1 PRESSURE

In paragraph 1.12 we saw that loadings are often applied to structures in the form of pressures. Quite often the applied force, instead of being concentrated at a single point, is *distributed* over a finite area.

When a force is distributed over part of the external surface of a structure, it is referred to as a *pressure*. Pressure is defined as force per unit area:

$$\text{Pressure (P)} = \frac{\text{total external force}}{\text{surface area over which force acts}}$$

In the study of structural behaviour, we use the term pressure when we are discussing *forces at the interface between two materials*. The pressure of water against the wall or floor of a pool, the pressure of the wind on the wall of a building, the pressure between a heap of sand and the floor — these are all common examples of pressures applied to the surface of some object (Figure 2.1). We will use the term pressure when we are referring to an external force distributed over an area at the surface of an object.

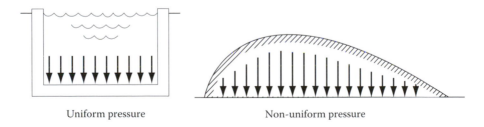

Uniform pressure   Non-uniform pressure

FIGURE 2.1

EXAMPLE 2.1

A brickwork pier carries a vertical axial load of 60 kN and is supported on a concrete footing. What are the stresses within the brickwork, and the pressures on the upper and lower surfaces of the footing?

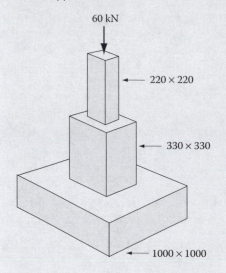

60 kN

220 × 220

330 × 330

1000 × 1000

The upper part of the pier carries a compressive force of 60 kN, and this force is distributed over an internal area of $220 \times 220 = 48\,400$ mm². Thus the compressive stress in the upper part of the pier:

$$= \frac{60 \times 10^3}{48\,400} \text{ N/mm}^2$$

$$= 1.24 \text{ MPa}$$

The lower part of the pier has an area of 330×330 mm², and the stress will be:

$$= \frac{60 \times 10^3}{330 \times 330} \text{ N/mm}^2$$

$$= 0.55 \text{ MPa}$$

The *pressure* exerted by the brickwork on the top surface of the footing will also be 0.55 MPa, because the same force of 60 kN is being applied over an external area of footing 330 mm square. By convention, however, we express a pressure in kPa.

$$\therefore \text{ Pressure on top of footing} = 550 \text{ kPa}$$

The pressure exerted on the bottom of the footing by the soil results from a force of 60 kN acting on an area 1 m square.

$$\therefore \text{ Pressure on bottom of footing} = \frac{60 \text{ kN}}{1 \text{ m}^2} = 60 \text{ kPa}$$

(In this example, for simplicity, we have ignored the self-weight of the brickwork and concrete.)

CALCULATION OF STRESS

In calculations involving stress where the stress is to be expressed in MPa:

- express forces in Newtons, and
- express areas in mm².

The basic unit of pressure is the Pascal (Pa); a Pascal is the pressure produced when a force of one Newton is distributed over an area of one square metre.

$$1 \, \text{Pa} = \frac{1 \, \text{N}}{1 \, \text{m}^2}$$

The pressures involved in buildings often have magnitudes of several thousand Pascals and are usually specified in kilopascals (kPa); a kilopascal is a force of one kilonewton distributed over an area of one square metre.

2.2 STRESS

We saw in paragraph 1.13 that, in addition to the external forces acting on the surface, there are also *internal forces* acting within our structures. These internal forces are usually not localised at a point but are distributed over some *internal area*. We could refer to these distributed internal forces as pressures, but instead we use the term *stress*. Both pressure and stress are defined as force per unit area, but, whereas pressure refers to external forces and surface areas, the term stress is restricted to internal forces and internal areas.

$$\text{Stress} = \frac{\text{total internal force}}{\text{internal area over which force acts}}$$

The units of stress are therefore the same as the units of pressure (i.e. Pascals), except that because stresses are commonly much larger than pressures, it is usual to record stresses in megapascals (MPa = 10^6 Pascals) (Figure 2.2).

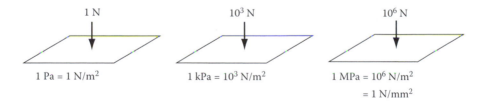

1 N	10^3 N	10^6 N
1 Pa = 1 N/m²	1 kPa = 10^3 N/m²	1 MPa = 10^6 N/m²
		= 1 N/mm²

FIGURE 2.2

You will often find that European texts refer to stresses in units of N/mm². *A megapascal is equal to one N/mm²*. You can check this for yourself. This produces the useful consequence that, provided we express forces in Newtons and areas in square millimetres, stresses will automatically be expressed in megapascals.

BUCKLING OF LONG STRUTS

If you press downwards on the top of a long, straight, slender piece of *balsa-wood*, you will find that initially the strut appears unaffected by the force. (In fact, of course, the force causes stresses and strains, but your eye is unable to detect them.) As you increase the force, the strut will quite suddenly buckle sideways and probably break. The maximum force that can be applied to a strut before buckling is called the critical buckling load. The phenomenon of buckling was first studied by the mathematician *Euler*, and the formula he derived is still the basis of our methods of column design.

For a strut of length L with a square cross-section of width D, and made of a material with a Young's Modulus of E, the critical buckling load is given by:

$$P_{cr} = \frac{\pi^2 E D^4}{12L^2}$$

$$\text{i.e. } P_{cr} = \frac{\pi^2 E}{12(L/D)^2} \times (\text{area})$$

Two important characteristics of columns may be observed from this equation:

(a) The load that a column can carry varies inversely with the square of the *slenderness* of the column, where 'slenderness' can be defined as the ratio of length to width. If the length of a column is doubled, the column will carry only one quarter of the load before it buckles. This is a most important fact: long columns are very inefficient structural members because they have to be made very thick to avoid buckling.

(b) The buckling load of a column does not depend upon the strength of the material of which it is made, but depends upon the Young's Modulus of the material. If two identical columns are made of two materials, each having the same modulus of elasticity, they will both buckle at the same stress even though one material may be ten or a hundred times stronger than the other.

Euler's theorem is true only for 'pin-ended' struts, i.e. struts that are free to rotate at both ends. If a strut is clamped at both ends, the tendency to buckle is greatly reduced, and the load that the column can safely carry will be increased about fourfold.

The distinction between stress and pressure is made quite clear in Example 2.1. When we are dealing with a force applied to the surface of the material (as, for example, the force applied by the pier to the footing), we speak of 'pressure *on* the surface' and usually use units of kPa. When the force is internal, we speak of the 'stress *within* the material' (e.g. the stress in the brickwork) and express that stress in MPa.

2.3 TYPES OF STRESS

Referring again to paragraph 1.13, we recall that there are only two possible kinds of internal forces: *normal forces* which act perpendicularly to a surface, and *shear forces* which act parallel to the surface. Consequently, there will be two types of internal stress, normal stress and shear stress (Figure 2.3).

Normal or axial stresses are produced by forces that tend to make an object *lengthen or shorten*. When we pull on a piece of string, or squeeze a pencil between our fingers, we are producing normal stresses in the string and the pencil. Normal stresses are either tensile (if the material lengthens) or compressive (if it shortens).

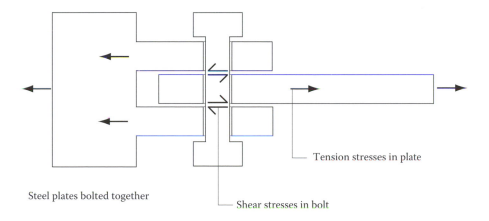

Tension stresses in plate

Steel plates bolted together

Shear stresses in bolt

FIGURE 2.3

Shear stresses are produced by forces that tend to make certain parts of the object *slide* over other parts. If this sliding is prevented, the shear stresses produce *angular distortions* within the object. When we grasp a rope and pull it, we produce shear stresses in the skin of our hands, and in the surface layers of the rope. Slide the ball of your thumb across the ball of your middle finger; shear stresses are produced in your skin. If you increase the pressure between thumb and finger you can see the angular distortions that are produced by shear if sliding is resisted.

EXAMPLE 2.2

A straight steel bar is 40 mm square and 2 m long. The steel has a tensile strength of 400 MPa and a Young's Modulus of 200×10^3 MPa. Compare the ultimate strength of this bar when used as:

(a) a tie, and
(b) a strut.

As a tie, the maximum tensile force that could be transmitted is:

$$\text{Tension force} = \text{stress} \times \text{area}$$
$$= 400 \times 40^2 \text{ N}$$
$$= 640\,000 \text{ N}$$
$$= 640 \text{ kN}$$

As a strut, the maximum compressive force will not be greater than the force that just causes buckling, i.e.

$$\text{Compressive force} = \frac{\pi^2 E}{12(L/D)^2} \times \text{area}$$
$$= \frac{\pi^2 \times 200 \times 10^3}{12(2000/40)^2} \times 40^2 \text{ N}$$
$$= 105\,275 \text{ N}$$
$$= 105 \text{ kN}$$

Thus the steel bar is at least six times stronger in tension than it is in compression.

EXAMPLE 2.3

A steel plate 200 mm × 300 mm × 5 mm is glued to the face of a concrete wall, and a mass of 360 kg is hung from it. What stress is produced in the glue?

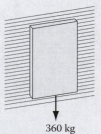

360 kg

The vertical force is parallel to the glue surface, and will produce *shear stress*.

$$\text{Force} = 360 \times 10 = 3600 \text{ N}$$
$$\text{Area} = 200 \times 300 = 60 \times 10^3 \text{ mm}^2$$
$$\therefore \text{Stress} = \frac{3600}{60\,000} \text{ N/mm}^2$$
$$= 0.06 \text{ MPa}$$

Therefore we see that normal stresses cause lengthening or shortening of the object, whereas shear stresses cause angular distortions. *Normal stresses* are computed by dividing a force by an area *perpendicular* to the force. For *shear stresses* we use an area *parallel* to the direction of the force.

2.4   STRAIN AND ELASTICITY

When external forces produce stresses within a body, the body also undergoes a change in its shape. If shear stresses are produced, angular changes will occur. If normal stresses are produced, the body will change its dimensions by lengthening in tension or shortening in compression. The change in dimension, expressed as a fraction of the original unstressed dimension, is called the *strain*.

$$\text{Strain} = \frac{\text{change in length}}{\text{original length}}$$

For many materials, stress is proportional to strain. That is, if stress is plotted against strain, the resulting graph is a straight line. The constant of proportionality between stress and strain is called Young's Modulus of Elasticity (E).

$$\text{Stress} = E \times \text{Strain}$$

Young's Modulus is used as a measure of the stiffness of a material. Those materials that have a large value of E can be used to form efficient, stiff structures.

2.5   DESIGN OF TENSION ELEMENTS (TIES)

Materials differ greatly in their ability to resist tensile stress. The tensile strength of concrete is about 4 MPa, and that of structural steel about 400 MPa. Thus a steel bar would not break until the tensile stress reached 400 MPa, i.e. a bar 10 mm square would not break until the external force reached $400 \times 100 = 40\,000$ N.

If, however, we wished to build a structure in which one component was required to resist a tensile force of 40 000 N, we would be very unwise to use a steel bar only 10 mm square, because this bar would be just on the point of breaking each time it was required to carry the load. The component would have no margin of safety against accidental overload, or against errors in the manufacture of the steel itself or of the component. The first requirement of a structure is safety. Every structure is designed to have a reserve of strength over and above that required for the applied loads.

WORKSHEET 2.1

2.1 Figure 1 shows one face of a braced cube. If a force of 2 kN is applied horizontally at joint B, and if all the steel bars AB, AC, BC and BD are connected by hinges or pins at A, B, C and D:

    (a) What is the magnitude of the horizontal reactions at D and C?
    (b) Consider the equilibrium of the rigid triangle ABC, and calculate the force in member BD. Is BD a tie or a strut?
    (c) Is there a vertical reaction at C?
    (d) Consider the equilibrium of the pin at C, and calculate the forces in members CA and CB. Are these members ties or struts?

2.2 Figure 2 shows a beam AB supported by a column AC and a cable BD.

    (a) What is the tension in the cable BD? If this cable has a diameter of 10 mm, what is the tensile stress?
    (b) What force is transmitted at A from the post AC to the beam AB? If this force is transmitted by a 10 mm diameter bolt as shown in Figure 2.3, what is the shear stress in the bolt? (Note that there are *two* surfaces resisting shear in this bolt.)
    (c) What are the reactions at C?

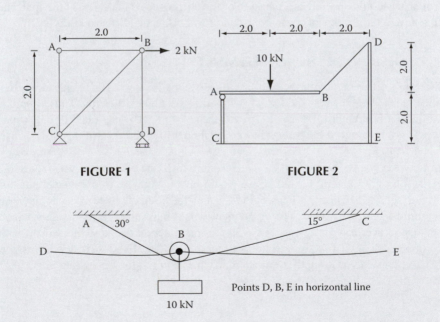

FIGURE 1   **FIGURE 2**

FIGURE 3

This reserve of strength is provided by the method used in the design of the structure. There are two basic design methods available which appear to differ greatly in their approach, but which in fact lead to fairly similar results.

In *Ultimate Strength Design* our concern centres on the strength of the structure at failure. We are seeking to ensure that there is an adequate margin between the load at which the structure might fail and the load it will actually be asked to carry. In this method, we increase the design loads by a factor called the *Load Factor*, and then ensure that the structure is just able to carry those increased loads without failure or collapse. For example, we might apply Load Factor of 1.6 to a tension member that was to carry a force of 100 kN. We would then make the member of such a size that it could carry a force of 160 kN before it failed in tension.

The other main design method is called the *Permissible Stress Design* method. The philosophy of this approach is based upon the following:

(a) The material of which the structure is composed is assumed to behave linearly elastically (stress is proportional to strain);
(b) For each of the possible forms of stress the material has a *permissible or allowable stress*, i.e. a stress that the material can sustain with an adequate margin of safety;
(c) The structure is proportioned so that *no part* of it is required to sustain a stress higher than the permissible stress when the structure is supporting its working loads.

Permissible stresses themselves are derived from the concept that any material will eventually reach a stress at which it ceases to be useful structurally. This 'useful' stress, for example, could be the yield stress for steel, or the ultimate compressive stress for concrete. The permissible stress is always less than this useful stress by some factor judged necessary to provide sufficient safety for the structure.

$$\text{Permissible stress} = \frac{\text{useful stress}}{\text{stress safety factor}}$$

For example, the yield stress of a particular type of steel may be 270 MPa. With a safety factor of 1.6, the permissible stress in tension would be $270/1.6 = 169$ MPa. A particular beam in which the *maximum* tensile stress anywhere in the beam was, say, 150 MPa would be judged acceptable, whereas a beam that contained a stress of 180 MPa would be unsafe, and would need to be strengthened.

Codes of Practice typically recommend ultimate strength capacities for the Ultimate Strength Design method, or permissible strengths for given materials where the permissible strength approach is still covered by the Code (see Section 9.6).

2.6 TIES AND STRUTS

A structural member that carries an axial tensile stress is called a *tie*. Ties are extremely efficient structural members, because every part of the material is fully

WORKSHEET 2.1   (CONTINUED)

2.3 A square timber column has dimensions 100×100 and a Young's Modulus of 10.5×10^3 MPa.

(a) What is the buckling load of this column if it is pin-ended with a height of 3 m?
(b) If the column is required to have a Safety Factor against buckling of 2.5, what would be the *safe* load?
(c) If the column height is increased to 6 m, what is the safe load?
(d) What is the safe load of a column 200×200 with a pin-ended height of 3 m?

2.4 In Figure 3, we are using a 'flying-fox' to transport a concrete beam from one side of a building site to the other. The beam hangs from a wheel at B, and two ropes BD and BE can be used to pull the beam in either direction. One of these ropes will always be slack.

(a) What are the forces in AB, BC, BD and BE with the beam stationary in the position shown?
(b) How must these forces be altered to start the beam moving to the right?

utilised. Consequently, we find that very large structures, such as bridges and sports stadia, often make extensive use of tensile members, most often in the forms of cables. Figure 2.4 shows some typical uses of cable structures.

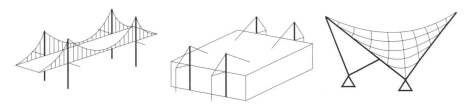

FIGURE 2.4

A structural member that carries an axial compressive stress is called a *strut* or a *column*. The behaviour of struts when stresses are large is far more complicated than that of ties, and depends greatly upon the shape of the cross-section, as well as length, and conditions at the end of the strut.

At one extreme, short sturdy columns usually fail by gradual crushing of the material at quite high values of stress. Brickwork piers would usually fall into this category. At the other extreme, long, slender columns usually fail by buckling. That is, even though the loading is initially axial, the member suddenly bends side-ways like a beam and will break if the load is maintained.

The behaviour of structural components in tension, then, is very different to that in compression. Tensile members can be long and slender; a small amount of material can carry a great deal of force. Compressive members should be kept short: long members are inefficient, and can carry only small loads unless they are made very thick and heavy.

APPLICATION: TIES AND STRUTS, IL GRANDE BIGO
(SEE PHOTOGRAPHS OVER PAGE)

Built by the City of Genoa to celebrate the voyage of Christopher Columbus, 'Il Grande Bigo' (the big ship-crane) has two tall compression booms providing cable support to the tubular arches carrying the fabric roof, and another boom for suspending the cable-car elevator. All booms are of tubular steel, with the cigar-shape achieved by a series of rolled conic sections with gradually increasing taper, butt-welded together. This shape maximises cross-section width (diameter) at boom mid-length, where the buckling tendency is greatest, and so minimises the boom's slenderness ratio. This in turn maximises its compression load capacity, for a given tonnage of steel plate.

Project: Il Grande Bigo, Genoa, Italy; Architect: Renzo Piano Building Workshop, Genoa; Structural Engineer: Arup/Sidercad SA; Photos: Arup (Alistair Lenczner)

APPLICATION: STRUTS AND TIES, KURILPA BRIDGE
(SEE PHOTOGRAPHS OVER PAGE)

The Kurilpa pedestrian and cycle bridge in Brisbane, Australia, is a very large scale application of 'tensegrity' (tensional integrity), a structural idea developed by Kenneth Snelson and Buckminster Fuller in the 1940's and '50's. Major and minor masts (steel tubes up to 30 m × 905 mm × 19.1 mm wall thickness) are offset from the perpendicular both longitudinally and transversely. The 'flying spars' are pure tensegrity elements and introduce further apparent randomness into the structure. The main cables are spiral wound galvanised wire rope 30–80 mm diameter. The cable system serves to suspend the deck and canopy, stabilise the masts, and resist twisting and lateral movements of the superstructure.

Wind tunnel studies were carried out to check the system's aeroelastic stability. Three 3.2 tonne tuned mass dampers are suspended under the main span of the deck to control the risk of 'synchronous lateral excitation' from large crowds of pedestrians. Fittingly, the bridge provides a major access way to the Queensland Gallery of Modern Art, on the Brisbane South Bank.

Project: Kurilpa Bridge, Brisbane, Queensland, Australia; Architect: Cox Rayner Architects; Structural Engineer: Arup; Contractor: Baulderstone; Photos: David Sandison

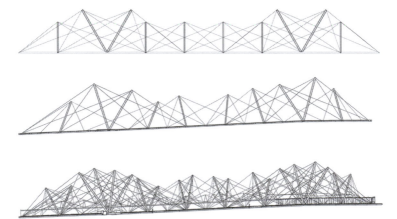

3 Loadings on Buildings

EXAMPLE 3.1

A floor for a cottage consists of hardwood floor boards 20 mm thick fixed to a series of parallel softwood joists that are 50 mm wide and 200 mm deep. The joists are spaced at 0.6 m between centres and their ends rest on walls 4.0 m apart. For what load should each joist be designed?

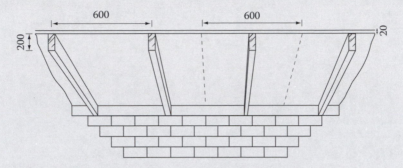

Because the joists are spaced 0.6 m apart, each joist carries a strip of floor 0.6 m wide and 4.0 m long. Thus, the load applied to *each metre* of each joist is the load on an area of floor 0.6 m × 1.0 m = 0.6 m².

Live Load

	Live load for domestic buildings	= 1.5 kN/m²
∴	Live load per metre of joist	= 1.5 × 0.6 = 0.9 kN/m.

Dead Loads

Density of floor boards	= 1000 kg/m³
Volume of flooring /metre of joist	= 0.6 m² × 0.020 m
∴ Weight of flooring/metre	= 0.6 × 0.020 × 1000 × 10 = 120 N/m
Density of floor joist	= 500 kg/m³
Volume of joist/metre	= 0.05 × 0.2 × 1.0 m³
∴ Weight of joist/metre	= 0.05 × 0.2 × 500 × 10 = 50 N/m
∴ Total Dead Load per metre	= 170 N/m = 0.170 kN/m
∴ Total Dead Load + Live Load	=1.07 kN/m

CONTENT OF CHAPTER 3

This chapter deals with the loads that may be applied to building structures, and with the way the loading may be calculated for any particular structural component. Some loads in nature are easier to compute than others: gravity loads are better understood than earthquake loads for example, but all need estimating numerically if we are to calculate and design safe structures.

Load types introduced in the chapter include "dead" and "live" loads, and more transient loads like wind, earthquake, impact and vibration.

3.1 LOADS

A building must be designed to withstand safely the most severe combination of forces or loads likely to be applied during its lifetime. The loads that are to be assumed when designing a structure are usually specified in 'loading codes'. The three types of load most commonly encountered are called Dead Loads, Live Loads and Wind Loads, although other loads such as impact, snow, thermal and earthquake may sometimes be important.

3.2 DEAD LOADS

Dead Loads are the loads that will be applied continuously during the life of the building. They are the gravity forces resulting from the mass of the structure, finishes,

---

TABLE 3.1
Typical Surface Densities of Materials and Constructions

Material or Construction	Weight/Unit Area kN/m^2
Ceiling, Gypsum plaster 13 mm thick	0.22
Roof, Corrugated galvanized steel	0.08
Roof, Terracotta tiles	0.59
Brick walls, per 25 mm of thickness	0.50
Concrete slabs, per 25 mm of thickness	0.60
Glass, per 25 mm of thickness	0.69

EXAMPLE 3.2

A small building in a fairly sheltered location has a flat roof having the form of construction sketched below. The roof structure consists of 50×200 softwood rafters, spaced at 0.8 m, supported on the external walls and on a central purlin.

Determine appropriate design loads for the various components.

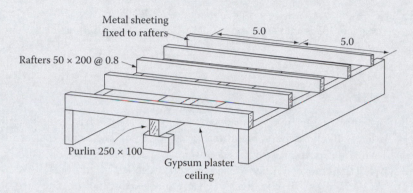

The following unit loads will be assumed:

Roof sheeting	$(7.8 \text{ kg/m}^2) = 0.078 \text{ kN/m}^2$
Softwood	$(500 \text{ kg/m}^3) = 5.0 \text{ kN/m}^3$
Plaster ceiling	$(22 \text{ kg/m}^2) = 0.22 \text{ kN/m}^2$
Maintenance live load	$= 0.25 \text{ kN/m}^2$
Wind uplift	$= 0.5 \text{ kPa} (= 0.5 \text{ kN/m}^2)$

(i) The roof sheeting is subjected to an uplift of 0.5 kN/m², but its own dead load is only 0.078 kN/m². Consequently, the sheeting will need to be fixed to the rafters with fixings capable of resisting a net upward pressure of 0.422 kN per square metre of roof.

(ii) For the rafters, we will compute the load applied *per metre length* of rafters. The spacing of the rafters is 0.8 m.

Dead load

Sheeting	$0.8 \times 1.0 \times 0.078$	$= 0.062 \text{ kN/m}$
Ceiling	$0.8 \times 1.0 \times 0.22$	$= 0.176 \text{ kN/m}$
Rafter	$0.20 \times 0.05 \times 1.0 \times 5.0$	$= 0.05 \text{ kN/m}$
	Total	0.288 kN/m

cladding and all other permanent components of the building. They are the most accurately estimated of all loads, since they may be calculated from the known densities of the materials used in construction.

Some typical and approximate material densities are:

Concrete	2400 kg/m^3
Brickwork	2000 kg/m^3
Sandstone	2300 kg/m^3
Hardwood	1000 kg/m^3
Softwood	500 kg/m^3

Many building materials are used in standardised thickness (e.g. sheet steel, wallboard, brickwork, etc.), and for these it is often more convenient to use the weight per unit surface area of construction. For example, 6.0 mm thick fibre-cement sheets have a weight of about 108 N/m^2; a sheet 1.0 wide by 3.0 m long would have a weight of $1.0 \times 3.0 \times 108 = 324$ N. Some values for other materials are listed in Table 3.1.

3.3   LIVE LOADS

Live Loads are the loads that are assumed to be applied to the floor of a building as a result of the activities carried on within the building. Consequently, they result from the type of *occupancy* of a particular part of a building, and are usually expressed in kN per square metre of floor area. The loading per square metre produced by students in a lecture-room, for example, is much larger than the loading in a domestic

TABLE 3.2
Typical Floor Live Loads

Classification	Uniformly Distributed Load kN/m^2
Public halls and theatres	4.0
Domestic buildings	1.5
Car parks	3.0
Grandstands	4.0
Library stack rooms	3.3 for each clear metre of height of room
Offices	3.0
Retail shops	5.0
Lecture theatres	3.0
Roofs, minimum maintenance load	0.25

EXAMPLE 3.2   (CONTINUED)

Live Load

$$0.8 \times 1.0 \times 0.25 = 0.20 \text{ kN/m}$$

Wind Load

$$0.8 \times 1.0 \times 0.5 = 0.40 \text{ kN/m}$$

The maximum *downward* loading will occur under dead load plus live load, with no wind load, and will be 0.488 kN/m. The maximum *upward* loading will occur when the live load is zero, and will produce an upward force of 0.112 kN/m.

(iii) For the purlin, we will again compute the uniformly distributed load per metre length. In doing this, we are assuming that the series of small equidistant point loads being applied by the rafters will produce the same effect on the purlin as would a uniform loading.

At each point where a rafter is supported by the purlin, the total load on two 2.5 m lengths of rafter will be transferred to the purlin. Consequently, the purlin will receive a series of concentrated loads, each 0.8 m apart. For dead load plus live load, each of these concentrated loads will be 5 × 0.488 kN, and the equivalent uniformly distributed load on the purlin will be:

$$\frac{5 \times 0.488}{0.8} = 3.05 \text{ kN/m}$$

The purlin self-weight will be:

$$0.10 \times 0.25 \times 1.0 \times 5.0 = 0.125 \text{ kN/m}$$

and the total downward load will be 3.175 kN/m.

For dead load plus wind load, we have seen that the loading on the rafters is an *upward* force of 0.112 kN/m so that the purlin loadings will be:

$$\frac{5 \times 0.112}{0.8} \text{ kN/m upward and 0.125 kN/m dead load}$$

i.e. 0.7 − 0.125 = 0.575 kN/m upward

Consequently, the design loadings for the purlin are 3.175 kN/m downward and 0.575 kN/m upward.

building, because there are more people per unit area of floor. The live load also includes the movable items of furniture and equipment normally associated with a particular type of occupancy. For example, a library is usually designed for a large live load in anticipation of the use of heavy book stacks. Some typical values of Live Load are listed in Table 3.2. There are minor variations between different countries' Codes of Practice.

3.4   WIND LOADS

When moving air impinges upon an object, it exerts a pressure that varies with the velocity of the air and with the shape and orientation of the object. The wind velocity to be assumed in design of buildings depends upon a number of factors, such as the geographical location, the type of surrounding terrain and the height above ground.

The effect of shape and orientation is allowed for by what are called 'pressure coefficients', applied to different parts of a building; the varying pressures result from the way in which the wind flows around the building (Figure 3.1). For example, the windward wall of a building will usually have a fairly large pressure coefficient associated with it, and must be designed for large pressures. The leeward wall, on the other hand, will have a negative coefficient, and hence will be subjected to a suction tending to pull the cladding away from the structure. A pitched roof may have pressure or suction depending upon the wind direction and the slope of the roof. The windward slope of steeply pitched roofs will be subjected to wind pressure, but low-pitched roofs and all leeward slopes will have suction producing an uplift on the roof.

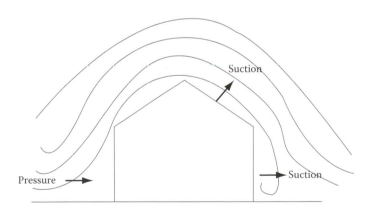

FIGURE 3.1

WORKSHEET 3.1

3.1 The diagram shows a timber garage for two cars. The tiled roof is sup-
    ported on 50×25 battens spaced at 0.3 m centres which are carried
    on 50×125 rafters at 0.6 m centres. The upper ends of the rafters are
    supported on a ridge-beam AB, which transmits its load to beam CD.
    Hardwood is used throughout. Neglect any effect that the pitch of the roof
    may have on the loads.

    (a) What is the uniformly distributed load on a typical square metre of
        roof? (Include live load in your calculations.)
    (b) Draw a sketch of a typical rafter, showing the load for which it must
        be designed.
    (c) Assume that the two beams AB and CD are each 300×75. For what
        loads should they be designed?

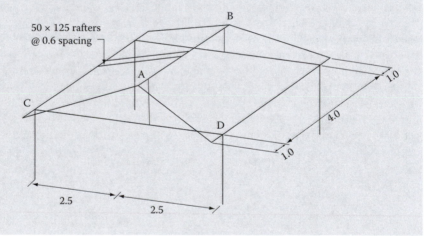

The selection of the correct wind load for a particular part of a building, then, is both critical and complicated, and we shall not attempt to deal with it fully here.

Even for small buildings, wind loads exert an important influence upon the choice and size of the structure. For tall buildings, the wind loads are often of overwhelming importance.

3.5 OTHER LOADS

Earthquakes occur in many parts of the world and buildings must be designed to withstand them safely. An earthquake causes violent shaking of the ground and can produce large horizontal forces on buildings. These forces depend mainly upon the severity of the shaking, the type of ground, and the mass and flexibility of the building. In areas in which they are rare or usually of low severity, it is often assumed that a structure designed to resist horizontal wind loads will probably be adequate in earthquakes as well. In areas of high earthquake risk, earthquake loads can often be so large that they determine the structure and even the form of the building.

Impact loads can result from moving objects, such as vehicles, that may strike a structure, and this possibility often needs to be considered during design. Vibrations caused by machinery or wind may pose a problem for some buildings. Changes of temperature or foundation settlement can also produce large forces in some structures. Horizontal earth pressure or water pressure or wave loading may be relevant. These and other loads may need consideration by the designer of any large building and decisions must be based upon an assessment of the risks involved.

3.6 CALCULATION OF LOADS

The calculation of the loadings acting upon a building structure is usually one of the first steps in analysing the structure. It can also be the most tedious, and one must adopt a systematic technique if errors are to be avoided.

Example 3.1 illustrates the general approach. For most structures we are dealing with linear members, and are therefore interested in a distributed load expressed in kN/m; we need to find out what load is supported by each *metre length* of the member. Usually, these linear elements are arranged as a parallel set (e.g. floor joists, rafters, concrete beams, etc.). Each element carries a strip of loading with a width equal to the spacing of the structural elements. Hence, the load per metre is the load applied to an area with a length of 1.0 m and a width equal to the spacing of the structural members (Figure 3.2).

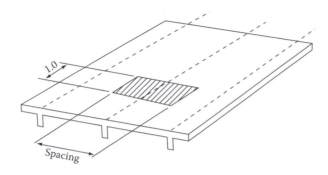

FIGURE 3.2

Each component of a building transmits loads to the components below; we can trace the path by which any given load is transmitted from its point of application to the foundations. Example 3.2 shows how we can calculate loadings progressively for a series of components, starting with the component that first receives the load, and moving on to each component in the chain.

Example 3.2 also brings out two important aspects of loading that deserve emphasis. First, because the loadings on a structure can result from several different causes, we often need to select *that particular combination* of dead load, live load or wind load that will result in the most adverse conditions for the structure. The dead load is always applied, but all or part of the live or wind loads may need to be omitted in order to predict the maximum design stresses at any particular point.

Secondly, the effect of wind loads is important, in that forces are often produced that are quite different from those we intuitively anticipate by considering dead load alone. This is especially so for the roofs of buildings. The uplift produced by wind quite often exceeds the downward dead load, and the roof structure needs to be securely tied down to the rest of the building fabric.

APPLICATION: LOADING, LIVE LOADS (SEE PHOTOGRAPHS OVER PAGE)

Providing support to applied loads is the fundamental purpose of most structures. Some of the loads are intentional, such as dead load of cladding and finishes, and live loads like occupants; moveable furniture, partitions and equipment; relocatable pipes and ducts. Some loads are attracted inadvertently, like wind, snow, and earthquake.

Shown here are some examples of different kinds of live loads specified by typical loading codes: open-plan office loading (Hewlett-Packard, Bracknell, UK); grandstand loading (Hong Kong Stadium, Rugby Sevens event); exhibition loading (Lille Grand Palais: Expo Exhibition Centre, Lille, France); plantroom loading (chiller room, No. 1 Poultry, City of London). Code recommendations need examining for particular projects, but typical values for these live load situations are 3 kPa, 5 kPa, 20 kPa, 7.5 kPa respectively.

Project: Hewlett-Packard, Bracknell, UK; Fitout Engineer: Arup; Photo: Arup (Peter Mackinven)

Project: Hong Kong Stadium, Hong Kong; Architect: HOK; Structural Engineer: Arup; Photo: Arup (Colin Wade)

Project: Lille Grand Palais: Expo Exhibition Centre, Lille, France; Architect: Rem Koolhaus/ Office for Metropolitan Architecture (OMA)/FM Delhay-Caille; Structural Engineer: Arup; Photo: OMA

Project: No. 1 Poultry, City of London; Architect: Michael Wilford and Partners; Structural Engineer: Arup; Photo: Arup (Peter Mackinven)

4 Graphical Statics

EXAMPLE 4.1

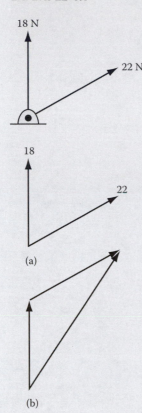

What resultant force do the two forces of 18 N and 22 N produce on the fixed eye-bolt?

(i)   Draw the Free-Body Diagram (a) showing the forces in their correct spatial relationship (i.e. meeting at a point).

(ii) Decide upon a suitable scale for the vector diagram (e.g. 1 mm : 1 N).

(iii) Draw a vector to represent the 18 N force (i.e. an arrow 18 mm long parallel to the 18 N force and of the same sense). Draw a vector to represent the 22 N force (Diagram b).

(iv) The resultant vector joins the tail of the first vector to the head of the second. Its length is 31.8 mm and it is inclined at 48° to the horizontal.

(v)   The resultant force has magnitude 31.8 N, passes through the point of concurrency and is inclined at 48° to the horizontal.

CONTENT OF CHAPTER 4

Analytical, or mathematical, methods are usually used to calculate the forces needed to hold structures in equilibrium, often with a computer. Graphical methods can be an interesting and quick alternative, however, and can help provide a better intuitive understanding of how the forces work together. In Chapter 1 we considered the analytical methods, and in this chapter we will examine some graphical methods.

The accuracy of the solutions obtained by graphical methods depends of course upon the accuracy of your draftsmanship. Manual graphic statics requires the same careful approach as any other manual graphical work in architecture: drawing instruments, a sharp pencil and an intelligent appreciation of the logic of what you are doing are all essential. Remember to state the *scale* of any diagram that you draw!

4.1 VECTORS

In graphical statics, we represent a force by a vector (Figure 4.1). A vector is shown graphically as an arrow, for which:

(a) the *length is proportional to the magnitude* of the force to a *stated scale*;
(b) the *direction is the same as the direction* of the force with respect to a fixed frame of reference;
(c) the *sense*, as indicated by the arrow head, *is the same as the sense* of the force.

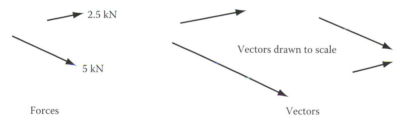

Forces

Vectors drawn to scale

Vectors

FIGURE 4.1

Note that a graphical vector *has no fixed location*; it differs in this respect from a force which must be specified by a point on its line of action. The value of graphical solutions lies in our ability to represent forces by vector arrows, and then arrange the vectors on the page *in any order or location we choose*. Thus, graphical solutions usually require two diagrams:

(a) a *Free-Body Diagram* or Space Diagram which shows all forces acting upon a body *in their correct spatial location*.
(b) a *Vector Diagram* or Force Diagram which shows the vectors by which the forces are represented.

APPLICATION OF GRAPHICAL METHODS

We have seen in Chapter 1 that the forces on any body in equilibrium must satisfy the three equations of equilibrium:

$$\Sigma F_x = 0 \quad \Sigma F_y = 0 \quad \Sigma M = 0$$

These three equations can be used to determine not more than three unknown force characteristics (magnitude, direction or location). The graphical solutions we are discussing in this chapter can do no more than is possible by the analytical approach on which they are based, and therefore cannot be used to determine more than three unknowns. If, in addition, the forces are concurrent, then only two unknown force characteristics can be determined, because the equation for moments does not produce a useful equation. It should be especially noted, therefore, that *for concurrent forces, two unknowns may be determined; for non-concurrent forces, three unknowns may be obtained.*

EXAMPLE 4.2

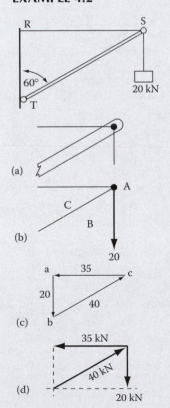

Determine the forces in the tie (i.e. tension member) RS and the strut (i.e. compression member) TS.

(a) It is essential to identify an *object* upon which the forces act. For example, imagine a bolt at S that attaches the load to the tie and to the strut.

(b) Draw the Free-Body Diagram for this bolt, and identify the forces by Bow's Notation. (The *sense* of the forces in tie and strut is unknown, and thus no arrowheads *need* be shown at this stage.)

(c) Represent the only known force AB by vector $\overline{ab}$ 20 mm long. The vector representing BC will lie on a line passing through b and parallel to BC. The vector representing CA will pass through a and be parallel to CA. The intersection of these two lines will determine point c. Complete the force diagram by showing the senses of the vectors and measuring their lengths.

(d) Complete the Free-Body Diagram by showing the magnitude and sense of the forces.

Both of these diagrams must be drawn accurately and to scale; the accuracy of graphical methods depends upon the precision with which these diagrams are drawn.

The addition of two vectors $\bar{A}$ and $\bar{B}$ is accomplished graphically by the triangle rule to produce the sum $\bar{R}$ (Figure 4.2). This may be extended to obtain the sum of n vectors, $\bar{A} + \bar{B} + \bar{C} + \cdots$ (Figure 4.3). It should be noted that the *order* in which the additions are made is immaterial: the sum $\bar{R}$ obtained from $\bar{A} + \bar{B} + \bar{C}$ is identical to the sum of $\bar{B} + \bar{C} + \bar{A}$.

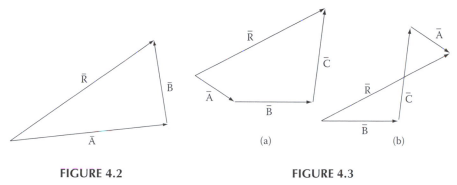

FIGURE 4.2   **FIGURE 4.3**

4.2 THE TRIANGLE OF FORCES: RESULTANTS AND EQUILIBRANTS

The Triangle of Forces is used for systems involving not more than *three concurrent forces*, and makes direct use of the rule for vector addition. Its application to two common types of problem is as follows:

(a) Where two forces are fully specified (magnitude, direction and location) and their *resultant* is required. The two known force vectors are drawn head-to-tail, and the resultant vector is obtained in magnitude and direction by joining the tail of the first vector to the head of the second (Figure 4.4). The resultant force has magnitude, direction and sense as given by the resultant vector, and is located at the point of concurrence of the two specified forces.

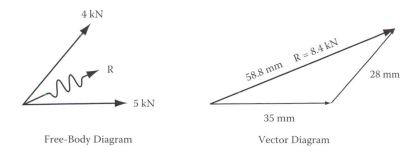

FIGURE 4.4

Principles of Structure

EXAMPLE 4.3

Determine the tension in the cable at P and the reaction at the hinge at Q.

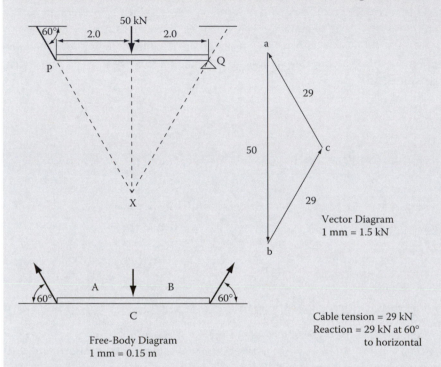

Vector Diagram
1 mm = 1.5 kN

Free-Body Diagram
1 mm = 0.15 m

Cable tension = 29 kN
Reaction = 29 kN at 60°
to horizontal

Because there are only three forces acting, the force at the hinge must pass through the intersection point of the other two forces. Hence, the Vector Diagram can be drawn.

(b) Where two forces are fully specified and a third force (the equilibrant) needed to maintain equilibrium is required. Since the equilibrant is a force that exactly opposes (i.e. is the reaction to) the resultant, the graphical construction is the same as for the previous case, but the direction of the closing vector is reversed (Figure 4.5). Thus, *if the forces are in equilibrium, the vector arrows run head-to-tail around a closed diagram.*

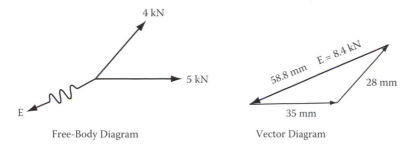

Free-Body Diagram   Vector Diagram

FIGURE 4.5

4.3   BOW'S NOTATION

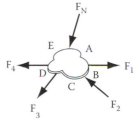

FIGURE 4.6

When dealing with systems of forces in equilibrium, it is often convenient to identify forces and vectors by a nomenclature known as Bow's Notation. In the Free-Body Diagram, capital letters are assigned to the spaces between the forces; the force is then named by the spaces on either side of it. The force F_1 is called AB, the force F_2 is called BC. *The usual convention is to name the forces by reading spaces clockwise around the Free-Body Diagram* (Figure 4.6).

In the Vector Diagram, the vectors are named by *lower case letters*, with the earlier letter of the alphabet being attached to the tail of the vector, and the later letter to the head of the vector. (This results from our convention of reading spaces *clockwise* around the Free-Body Diagram.) Thus force F_1 (or AB) is represented by the vector a → b, and Force F_4 by vector e ← d.

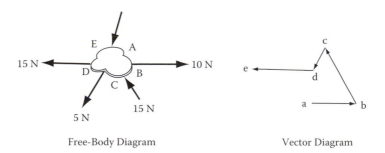

Free-Body Diagram   Vector Diagram

FIGURE 4.7

EXAMPLE 4.4

Determine the tension in the cable at P and the reaction at the hinge at Q.

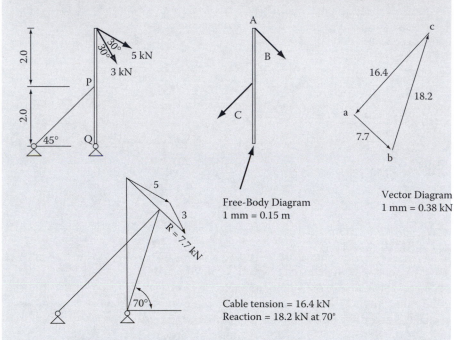

Free-Body Diagram
1 mm = 0.15 m

Vector Diagram
1 mm = 0.38 kN

Cable tension = 16.4 kN
Reaction = 18.2 kN at 70°

By finding the resultant of the two applied forces of 5 kN and 3 kN, the problem can be turned into one involving only three forces. The solution is then similar to Example 4.3.

Consequently, the two diagrams would appear in Bow's Notation as shown in Figure 4.7.

4.4 THE TRIANGLE OF FORCES: TWO UNKNOWN FORCES

Frequently force systems are encountered in which only one force is fully specified, and one unknown (magnitude or direction) is required for each of the other two forces. Bow's Notation is especially helpful here. The known force vector is shown and the third point of the vector triangle is plotted from the information given. The procedure for two unknown magnitudes is shown in Figure 4.8. You should devise for yourself the procedure when two directions *or* one magnitude and one direction are unknown.

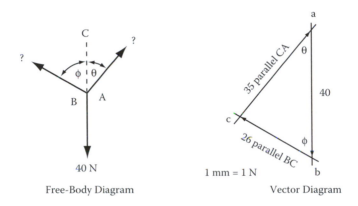

Free-Body Diagram   Vector Diagram

FIGURE 4.8

4.5 CONCURRENCY OF FORCES

If a body is acted upon by only three forces, the forces must be either concurrent or parallel (refer to paragraph 1.9). This fact is often useful in problems involving reactions where there are three unknown force components (Examples 4.3 and 4.4). The method of solution is to determine graphically the point of concurrency of two forces, which in turn fixes the direction of the third force. The magnitudes of the two unknown forces may then be found from a triangle of forces.

In Example 4.3, the direction of the reaction at the hinge is initially unknown. However, the point of intersection of the other two forces (x) can be determined graphically, and the reaction at the hinge must pass through this point. This determines the direction of the reaction at the hinge. In Example 4.4, the solution is obtained in two steps. First, the resultant of the two applied loads is determined (thereby reducing the problem to one involving only three forces), and secondly the point of intersection of this resultant and the force in the cable is obtained graphically. Note particularly the use of Bow's Notation in these two examples.

WORKSHEET 4.1

4.1 A builders' elevator (Figure 1) is stabilised by three guy-ropes attached to a block of concrete at ground level. The forces in the guy-ropes are as shown.

(a) What resultant force do the ropes exert upon the block?
(b) What volume of concrete would be required to keep the block in equilibrium in the vertical direction if the density of concrete is 2400 kg/m^3?

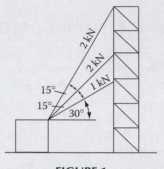

FIGURE 1

4.2 Wind produces a force of 400 N at the top of a flag pole shown in Figure 2. What is the force in the guy-rope and along the flag pole?

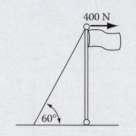

FIGURE 2

4.3 The roof of a stadium is a cable structure in which the cables apply a horizontal force of 300 kN to the abutments (Figure 3). What are the forces in the cable-stay AB and the strut AC?

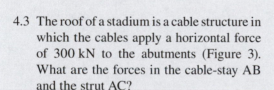

FIGURE 3

4.4 A load of 300 N is suspended from the bracket shown in Figure 4. Use the three-force condition (paragraph 1.9) to determine the force in AB and the force at C.

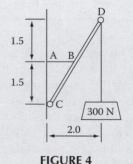

FIGURE 4

APPLICATION: GRAPHICAL STATICS, PATSCENTRE
(SEE PHOTOGRAPHS OVER PAGE)

Patscentre is a research facility for PA Technology, near Princeton, New Jersey, USA. During development of the roof structure geometry, graphical statics was used to demonstrate the greater efficiency of Option 2 over Option 1. In the latter, the resultant of tension forces in CD and CE must lie between CD and CE, so cannot enter shaded zone ABC. BC must therefore be in compression. Conversely, in Option 2 the resultant of CD and CE falls inside ABC, so ensuring tension in BC, and therefore allowing use of a light tie as preferred, rather than a visually bulkier strut.

While analysis by manual calculation and by manual graphical methods have been superseded by computer-based methods for routine analysis and design, they are still commonly used during scheme stage and at 'decision points' like the above. Graphical methods provide visible evidence of the interplay of forces and so reinforce understanding of behaviour.

Project: Patscentre, for PA Technology, Princeton, NJ; Architect: Richard Rogers and Partners/Kelbaugh and Lee; Structural Engineer: Arup/Robert Silman Assocs; Photo: Otto Baitz/Esto; Diagrams: Arup

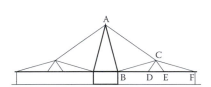

Option 1 geometry

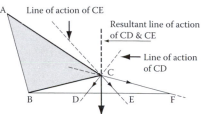

Option 1 force vectors

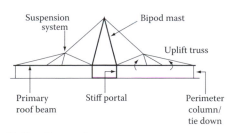

Option 2 geometry

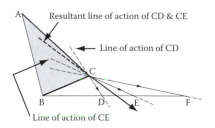

Option 2 force vectors

5 Pin-Jointed Structures

BEHAVIOUR OF A PIN-JOINTED MEMBER

The members used in the structures we are considering in this chapter are all straight and pin-ended. They are joined to the other parts of the structure by pin-joints, or hinges at their ends. If we draw a Free-Body Diagram of such a member, showing the two forces that can be applied at each hinge (refer to paragraph 1.10), we can deduce two important characteristics. First, by taking moments about hinge A, we see that the transverse force V_B must be zero. Thus *a pin-ended member is not subject to shear force*. Secondly, because both V_B and V_A are zero and because no bending moment can be applied by the hinge, *a pin-ended member is not subjected to bending moment*. Both these statements assume no extra load is applied to the member or between the pin ends.

Pin-ended members carry axial forces only; they are quite free of shear and bending.

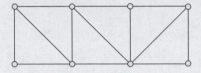 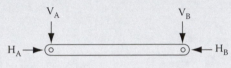

CONTENT OF CHAPTER 5

Pin-jointed structures are among the most efficient of man-made structures. They range in complexity from simple timber frames, to large two-dimensional bridge trusses, to immense space-frames capable of roofing huge areas.

In this chapter, we are concerned with two-dimensional pin-jointed frames, or trusses. We will consider three useful methods for analysing them to deduce the effects of externally applied loads: graphical methods; the method of joints; and the method of sections.

5.1 PIN-ENDED MEMBERS

Many framed structures (e.g. trusses, A-frames, wind bracing) are designed as pin-jointed frames. All the joints between members are assumed to consist of frictionless pins, incapable of transmitting a moment. Moreover, as a first approximation, all loads are assumed to be applied at the joints. Hence, if the members are all straight, they will all be quite free from bending. Each member will be subject to axial forces only, pure tension or pure compression. In most instances these assumptions are fairly close to the truth, and it is only for very large, important or unusual structures that a more sophisticated method of analysis is required.

There are three methods available for use in determining the forces in the members of a pin-jointed plane frame. Each of them depends for its validity upon the fact that if the frame as a whole is in equilibrium, then any isolated portion of it must be in equilibrium. The forces needed to maintain this portion in equilibrium may then be obtained from the equations of equilibrium.

5.2 GRAPHICAL METHOD

Graphical statics, introduced in Chapter 4, can aid intuitive understanding of forces. The same is true of graphical methods for determining the forces in pin-jointed frames, such as the Maxwell Diagram method. The Maxwell Diagram consists of a force diagram which contains within the one figure the force polygons for every joint in the framework.

To construct a Maxwell diagram, it is first necessary to determine the external reaction forces. This is most easily done by calculation (although graphical methods such as the link polygon can be used). All external forces are then identified by Bow's Notation, using letters of the alphabet in clockwise order around the frame. Identify

EXAMPLE 5.1

Use a Maxwell diagram to determine the forces in all members of the pin-jointed truss shown in the diagram.

The vertical reactions at P and S will each equal 10 kN; draw the lines ab, bc and ca to represent the external forces.

Consider joint P; the forces involved are A1, 1C and CA. The triangle ac1 establishes the point 1.

Consider joint Q, and plot point 2 from the forces A2 and 21 etc.

Measure c1, 1a etc., and show magnitude and direction of each force on a line diagram of the truss.

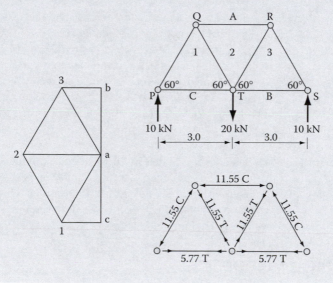

the closed spaces formed by the members of the frame by numerals 1, 2, 3 (Figure 5.1). Commence the Force diagram in the usual way by representing the external forces by vectors ab, bc, cd, etc. Now consider the joint at the left-hand reaction (P). The forces on the frictionless pin at this joint (proceeding *clockwise* around the joint) are A1, 1E and EA, and vectors could be drawn parallel to these forces in the usual manner to form a triangle of forces. Instead, draw lines (*without* arrow-heads) through point 'a' parallel to A1 and through 'e' parallel to 1E. The intersection of these lines determines point 1, the third point on the triangle of forces for this joint.

Next consider the joint adjacent to spaces A and B. The forces acting on this joint are 1A, AB, B2 and 21. Turning to the force diagram, trace the existing lines 1a, ab, and determine the location of point 2 such that b2 is parallel to force B2 and 21 is parallel to force 21. Proceed in turn with joints T, U, R, V, S and W, and plot the points 3, 4, 5 and 6.

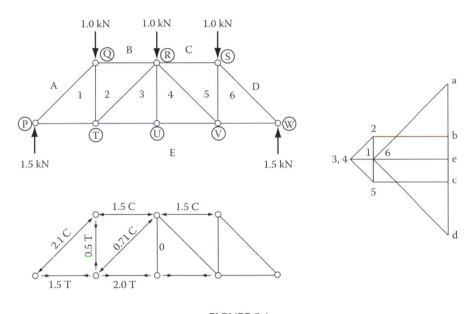

FIGURE 5.1

The lines drawn (e.g. 1e, 6a, 15, etc.) represent to scale the magnitudes and directions of the forces at the joints, e.g. member B2 is applying a force on joint Q equal in magnitude to the length of b2. To obtain the *sense* of each force at a joint, consider the forces in clockwise order around the joint and assign senses according to the order of the identifiers on the Force diagram. For example, for joint Q, vector b2 acts to the left, 21 is downward and 1a is upward. Show the senses by arrowheads *adjacent to each joint*. Finally, write the magnitude of each force adjacent to the member.

EXAMPLE 5.2

Determine the member forces in all members of the cantilever roof truss shown.

Represent the external forces by the lines ab, bc, cd, de, ef and fa.

Consider the joints in order to plot points 1, 2, 3, 4 and 5.

The resultant member forces are shown on the original diagram.

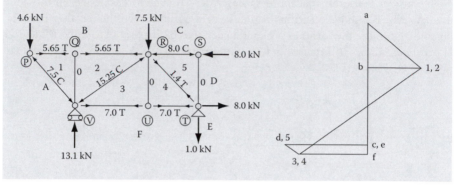

The following points in particular should be noted in this method:

(a) Usually the reactions must be obtained before commencing to draw the Maxwell Diagram.
(b) Forces at a joint are always considered *in clockwise order.*
(c) *Not more than two unknown forces can be determined at any one joint.* This fact will dictate the order in which the joints are examined, e.g. in the previous example, joint Q *must* precede joint T.
(d) The vectors in the Maxwell Diagram are drawn *without arrowheads.*
(e) Throughout the process, attention should be directed *to the joints*; all forces considered are applied to the joint not to the member.
(f) If two points in the Maxwell Diagram are coincident (e.g. 3 and 4 in Figure 5.1), no force is applied at the joint by the corresponding member.
(g) The results of the analysis should always be shown *on the diagram of the truss.* With the notation given, a member appearing thus $\longleftrightarrow$ is thrusting *towards* the joints. The member itself will be in compression. Similarly $\rightarrow\leftarrow$ indicates a tensional force in the member.
(h) The last line drawn on the Maxwell Diagram should pass through a previously established point. Usually, however, a graphical solution is not too precise, and a small 'triangle of error' will be formed by the last line. This should be indicated on the diagram.

5.3 METHOD OF JOINTS

There are two popular mathematical methods for finding the forces within truss members, both easily amenable to hand calculation. The first is called the 'method of joints'.

The procedure is very simple. At each joint, the forces are considered and two equations of equilibrium are established for force components in two mutually perpendicular directions. These equations are then solved simultaneously for up to two unknown forces.

For the truss shown in Figure 5.1, for example, we would first study joint P. The forces acting on the pin at P are shown in Figure 5.2. Because the reaction is upward, the force in member PQ will be downward. Because the force F_{PQ} has a horizontal component acting towards the left, the force F_{PT} will act towards the right. (Thus, the sense of the forces can often be predicted.) We may now write the two equations of equilibrium for the pin at P:

$$\text{From } \Sigma F_y = 0 \qquad + 1.5 - F_{PQ} \cos 45° = 0$$
$$\therefore \ F_{PQ} = +2.12 \text{ kN}$$
$$\text{From } \Sigma F_x = 0 \qquad + F_{PT} - F_{PQ} \cos 45° = 0$$
$$\therefore \ F_{PT} = 1.5 \text{ kN}$$

These two forces should now be marked on the truss diagram, together with the necessary forces of opposite sense at the other ends of members PQ and PT.

EXAMPLE 5.3

Use the method of joints to compute the member forces in the truss shown. Consider each joint in turn:

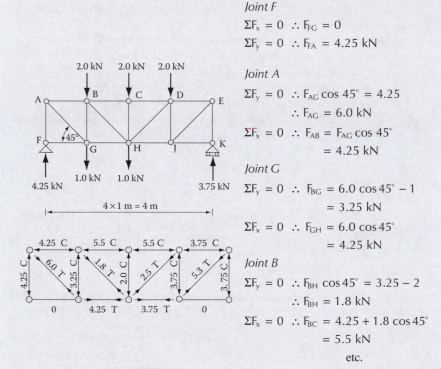

Joint F

$\Sigma F_x = 0$ ∴ $F_{FG} = 0$

$\Sigma F_y = 0$ ∴ $F_{FA} = 4.25$ kN

Joint A

$\Sigma F_y = 0$ ∴ $F_{AG} \cos 45° = 4.25$

∴ $F_{AG} = 6.0$ kN

$\Sigma F_x = 0$ ∴ $F_{AB} = F_{AG} \cos 45°$

$= 4.25$ kN

Joint G

$\Sigma F_y = 0$ ∴ $F_{BG} = 6.0 \cos 45° - 1$

$= 3.25$ kN

$\Sigma F_x = 0$ ∴ $F_{GH} = 6.0 \cos 45°$

$= 4.25$ kN

Joint B

$\Sigma F_y = 0$ ∴ $F_{BH} \cos 45° = 3.25 - 2$

∴ $F_{BH} = 1.8$ kN

$\Sigma F_x = 0$ ∴ $F_{BC} = 4.25 + 1.8 \cos 45°$

$= 5.5$ kN

etc.

GEOMETRY OF TRUSSES

The geometry or arrangement of members in a truss depends upon a number of factors, and it is difficult to set down general rules that will apply in all instances.

The envelope formed by the upper and lower chords is usually determined by the particular application and by span/depth ratios. For example, in trusses used for pitched or saw-tooth roofs, the slope of the upper chord is determined by the pitch required for the proposed roof cladding. The span/depth ratio

The space diagram for joint Q can now be sketched, again assuming a sense for each unknown force.

$$\text{From } \Sigma F_y = 0 \quad + F_{TQ} + 2.12 \cos 45° - 1 = 0$$

$$\therefore F_{TQ} = -0.5 \text{ kN}$$

(i.e. F_{TQ} has the *opposite* sense to that assumed.)

$$\text{From } \Sigma F_x = 0 \quad + 2.12 \cos 45° - F_{QR} = 0$$

$$\therefore F_{QR} = 1.5 \text{ kN}$$

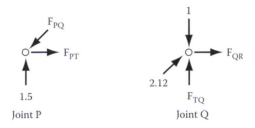

FIGURE 5.2

The following points summarise the method:

(a) Commence the analysis at a joint involving only two unknowns.
(b) Assign a sense to each unknown force at the joint. Usually the correct sense can be ascertained by inspection, but if not, arbitrarily assign a sense. When the force has been calculated a positive sign will indicate that the sense chosen was correct, and a negative sign indicates an incorrect choice, in which case the sense should be reversed before being indicated on the main truss diagram.
(c) As soon as each force is determined, it should be shown on the main diagram by an arrowhead positioned very close to the joint (see Example 5.3).

5.4   METHOD OF SECTIONS

The second mathematical method considers the equilibrium of a whole portion of a truss rather than just one joint at a time.

Suppose we wish to determine the force in member GH of the truss in Example 5.3. Imagine that the members BC, BH and GH are all cut through along a section a–a, and imagine that three forces, F_{BC}, F_{BH} and F_{GH}, are applied at the cut ends of the members to maintain the left-hand portion of the truss in equilibrium (Figure 5.3).

Since the left-hand part of the truss is now in equilibrium just as it was when jointed to the right-hand part, the three forces F_{BC}, F_{BH} and F_{GH} are identical with the forces

GEOMETRY OF TRUSSES (CONTINUED)

is chosen to provide a compromise between the structural economy that is obtained by using a relatively deep truss, and the amount of internal space that is wasted by the extra depth. It is greatly influenced also by the strength of the materials used (steel trusses can be shallower than timber ones) and by the amount of load to be carried. The depth of a truss is rarely less than one-twentieth of the span (e.g. for a very lightly loaded, closely spaced series of steel roof trusses) and may exceed one-eighth of span for heavier loads, or even a quarter of span for pitched or arched trusses.

The spacing of the web or diagonal members depends upon the truss depth and the pattern of loading. The diagonals, which provide the 'shear strength' of the truss, become inefficient if they are inclined at more than about 45° to the vertical. The pattern of loading is important for the two chords; it is preferable for loads to be applied *at* the joints, rather than *between* the joints. In a roof truss which supports purlins, for example, the spacing of joints along the upper chord is ideally related to the purlin spacing. Within these general requirements, one should attempt to keep web compression members as short as possible, so as to avoid the extra cost of long struts (refer to Chapter 2).

A number of frequently used truss forms are sketched below.

The Warren Truss is popular for steel parallel-chord trusses, since it permits a wide joint-spacing (in proportion to truss depth) along the chords.

The Pratt Truss, as a parallel-chord truss, is also popular, because the long diagonal members are ties and the short verticals are struts.

In the Howe Truss, it is the diagonals that are in compression; this form has been used for composite trusses with timber diagonals, and steel rods for verticals, for simpler joinery.

For trusses for pitched roofs, any of the above forms may be used, although the traditional Fink Truss has an advantage in that it makes good use of tension members, and is easily demountable.

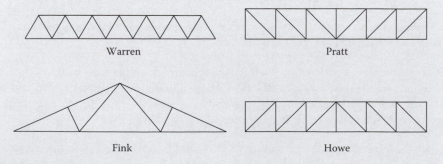

Warren   Pratt

Fink   Howe

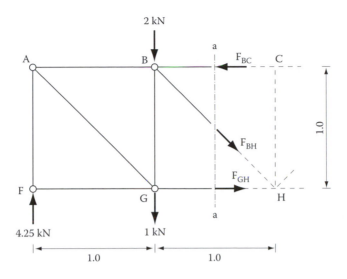

FIGURE 5.3

in members BC, BH and GH in the complete truss. The left-hand part must satisfy the equations of equilibrium, and the force F_{GH} may be found by taking moments about point B, the intersection of F_{BC} and F_{BH}.

$$\Sigma M_B = 0 \text{ (clockwise moments positive)}$$
$$= 4.25 \times 1 - F_{GH} \times 1$$
$$\therefore F_{GH} = 4.25 \text{ kN}$$

Similarly, for force F_{BC}, taking moments about H, the intersection of the other unknown forces:

$$\Sigma M_H = 0$$
$$= 4.25 \times 2 - 2 \times 1 - 1 \times 1 - F_{BC} \times 1$$
$$\therefore F_{BC} = \frac{8.5 - 3.0}{1.0} = 5.5 \text{ kN}$$

For force F_{BH}:

$$\Sigma F_y = 0$$
$$= 4.25 - 3 - F_{BH} \cos 45°$$
$$\therefore F_{BH} = \frac{1.25}{\cos 45}$$
$$= 1.8 \text{ kN}$$

WORKSHEET 5.1

Figures 1, 2 and 3 show three trusses:

(a) Determine the forces in all members of the trusses in Figures 1 and 3 by use of the Maxwell Diagram.
(b) Use the Method of Joints to determine the forces in the members of the trusses in Figures 2 and 3.
(c) Use the Method of Sections to determine the forces in members KL, KR and QR of the truss in Figure 2.

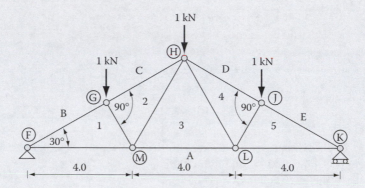

FIGURE 1

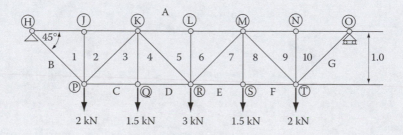

FIGURE 2

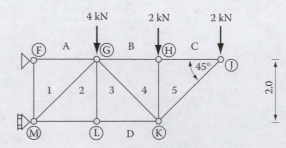

FIGURE 3

The following points summarise the method:

(a) Cut a section through a truss so as to cut three members for which forces are to be determined. Re-draw the part of the truss to one side of the section, and name the forces that are applied to the cut ends of the members.
(b) To find the force in an upper or lower chord member, take moments about the point of intersection of the other two forces.
(c) To find the force in an inclined member, use the equation for equilibrium of vertical forces.

5.5 SUMMARY

The three methods that have been discussed in this chapter all give a good insight into the behaviour of pin-jointed frames. The graphical method assists with visualisation of the forces inside the frame. The method of sections is often used when the forces in only a few of the members need to be obtained quickly. For example, for light-weight roof trusses, usually the two chords are made from continuous lengths of material, and only the maximum force in each chord need be determined. The method of joints is a general-purpose method; it provides a speedy solution for parallel-chord trusses, but can be slower for pitched trusses.

APPLICATION: ROOF TRUSSES, MENIL COLLECTION MUSEUM
(SEE PHOTOGRAPHS OVER PAGE)

An elegant example of 'pin-jointed' roof trusses occurs in the Menil Collection Museum in Houston, Texas. Secondary trusses span onto primary trusses which span onto columns. All trusses are parallel-chorded. The secondary trusses use as their bottom chords prefabricated ferro-cement 'leaves', which are carefully shaped to optimise the spread of diffuse daylight into the gallery space. The trusses are made up of tetrahedron-shaped elements, prefabricated in ductile cast iron. Cool air supplied through floor grilles, rises gently and is extracted at high level from beneath the glazed roof cladding that is supported by the trusses.

Project: Menil Collection Museum, Houston, Texas; Architect: Piano and Fitzgerald; Structural Engineer: Arup; Photos: (top and middle): Richard Bryant/Arcaid.co.uk; (bottom): Arup

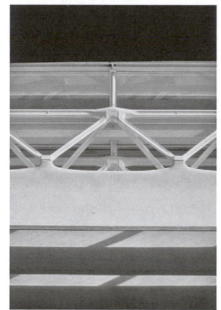

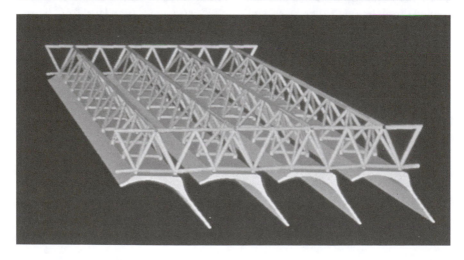

APPLICATION: PIN-JOINTED TRUSSES, RALEIGH-DURHAM
AIRPORT (SEE PHOTOGRAPHS OVER PAGE)

The roof structure for Terminal 2 at Raleigh-Durham International Airport, North Carolina USA, comprises 'king post' trusses with curved glued-laminated timber top chords, structural steel I-section king posts, and steel cable bottom chords. Dead and live loads are applied to the curved top chord via frequent longitudinal purlins. This means the top chord attracts bending moment and shear (Chapter 7) as well as compression, and so needs to be somewhat more substantial than the bottom chord which attracts only tension. Joints at the ends of the truss and at the bottom of the king post are 'pin joints'. The splice joint at mid-length of the top chord is not a pin joint — it can deliver compression, bending and shear. Full-size prototypes of the joint were tested at North Carolina State University to verify design assumptions.

Project: Raleigh-Durham International Air Terminal Building, North Carolina, USA; Architect: Fentress Bradburn Architects; Structural Engineer: Arup, with Stewart Engineering; Photos: (top) Jason A Knowles (copyright Fentress Architects), (middle left) Arup, (middle right and bottom) Stewart Engineering

6 Bracing of Buildings

TRIANGULATION

Consider a structure (A) composed of two bars pinned to each other and to their footings. We sense that the structure is quite stable and able to resist lateral loads. This type of structure is called a three-pinned arch or three-pinned frame. The bars may have any shape in elevation (B, C or D); the essential part is the presence of two rigid bars, three pins, and a rigid foundation to form the third leg of the triangle.

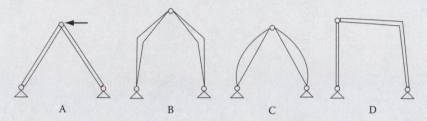

As soon as we add one extra bar (E), the structure becomes unstable; it is able to form a mechanism under lateral loading, and will collapse. There are many ways in which it can be stabilised, the simplest being to add another diagonal bar to *brace* the structure (F). The diagonal has restored the triangulated form which gave stability to the three-pinned arch (A), and will be in tension or compression depending upon whether the horizontal force is acting to the left or to the right (G, H).

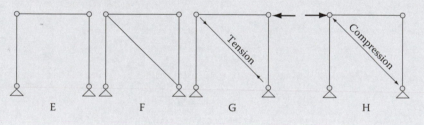

CONTENT OF CHAPTER 6

In this chapter we will consider the various ways in which a building may be braced so that it can maintain its shape under the action of horizontal loads. Methods include triangulation, bending frames, and shear walls, and they apply equally to multi-storey applications.

The emphasis is on comprehending the overall behaviour of the system, rather than on quantifying the loads and their effects. Other ideas introduced include loadpaths, torsional effects, and loadcase combinations.

6.1 VERTICAL AND LATERAL LOADS

All buildings must be able to withstand the appropriate loadings described in Chapter 3 without either overturning or being overstressed in any part. We may consider a building as being a rigid box, which rests upon the foundation (Figure 6.1). When vertical loads are applied, bending forces are set up in the horizontal components of the box and axial forces in the vertical components.

This vertical system of loading (e.g. dead loads and live loads) is the system we immediately envisage when we first come to select a structural form for a building. In fact, frequently a second system of loads, the lateral, or horizontal, loads, is more critical and more difficult to adequately withstand.

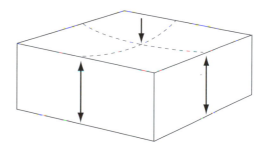

FIGURE 6.1

6.2 BASIC SYSTEMS

In the adjacent commentary, it will be seen that two-dimensional building frames are usually stabilised against lateral loads by the use of one of three structural systems:

(i) **Triangulation.** The system of posts and beams that transmit the vertical loads to the ground is braced by the addition of diagonal components that interconnect the various beam/column joints. All the joints may then be

RIGID FRAMES

Alternatively, we could stabilise the frame E by adding a knee-brace (J); this will triangulate the beam and the right-hand column and make them act as a single bar. Frame J then acts in a stable way similar to the three-hinged arch D. However a lateral load acting upon J will produce bending and bending moments in the beam and the column, and these bending moments may become quite large adjacent to the knee-brace (K). Frequently, two knee-braces are added, so that both columns assist in resisting the lateral load, and the bending moments are reduced (L). As the knee-brace becomes smaller, the bending moments become larger (even though the frame is still stable), the limiting case being reached when the joints themselves are made rigid (M). This type of structure is called a rigid frame or two-hinged portal; it resists lateral loads primarily by bending action.

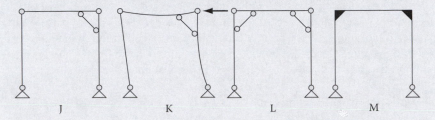

J   K   L   M

Notice that the legs of the two-hinged portal behave rather like cantilevers fixed at the top under lateral loading (N). In fact, of course, we could have a perfectly stable structure by having the columns firmly embedded in the foundation as true cantilevers. Structure P is like an inverted version of M. Of course, the lateral load could be resisted equally well by a single cantilever, and structures Q and R are of this type.

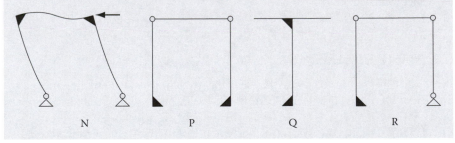

N   P   Q   R

pinned joints, which makes for easier on-site work, but the diagonal members may interrupt window openings or be regarded as unsightly. This system is usually the most efficient of all, because all the components are usually free of bending moments and carry only axial forces.

(ii) **Rigid Frames.** If the joints between column and beam (or footing) are not pinned but are made rigid, the structure is able to resist horizontal loads by bending of the columns and, usually, of the beams also. Large bending moments may exist in this type of structure. It is very difficult to make rigid joints in timber construction, but in reinforced concrete good rigid joints may be readily achieved because of the monolithic nature of the construction.

(iii) **Shear.** Where the supporting structure consists of walls, rather than beams and columns, it is logical to use these walls to resist lateral loads. When *the applied force lies in the same plane* as the wall, the wall is stressed in shear, and is called a Shear Wall. This system is the most economic of all, because it does not require the installation of any additional components.

6.3 BRACING OPTIONS

Let us now consider a simple three-dimensional building structure, consisting of two rigid frames and the section of wall cladding between them. We will assume that the cladding is fixed to horizontal beams (called *girts*) and that these are attached to the structure, as shown in Figure 6.2.

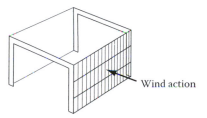

Wind action

FIGURE 6.2

For the building to withstand the wind pressure, the cladding must be strong enough to span as a beam between the girts, and the girts in turn must be strong enough to span between the frames. Obviously if the spacing of the frames becomes very large, the girts themselves will need to be made very strong, and for a long building it may be necessary to have a number of structural frames (Figure 6.3).

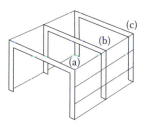

FIGURE 6.3

Now the wind load on the intermediate frame b is twice as large as that on the end frames (a or c), and therefore the intermediate frame would need to be made twice as strong. Often it is inconvenient or uneconomic to do this, so instead we may make the intermediate

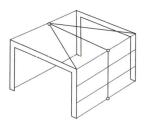

FIGURE 6.4

SHEAR

The single cantilever need not necessarily be a slender member as shown in Q. As it is made wider, its bending stresses become much smaller, until it becomes a wall, which transmits the lateral load to the foundations largely by shear. Such structures are called shear walls (S) and are commonly made from masonry or concrete.

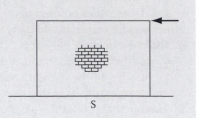

BRACING OF MULTI-STOREY BUILDINGS

So far we have been considering one-storey buildings only, but we may make a multi-storey building rigid simply by stacking rigid storey upon rigid storey (T, U and V).

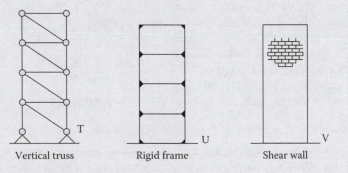

Vertical truss   Rigid frame   Shear wall

BRACING OF LONG BUILDINGS

It is often assumed that in a long building *every* column needs to be braced. This is not the case. Provided *one* column is braced strongly enough, all the other columns in that row can be held in position by simple horizontal members (e.g. girts).

frame a mechanism (e.g. structure E above), and support the tops of its columns by a *horizontal truss* in the plane of the roof as shown in Figure 6.4.

The intermediate column then acts as a vertical beam spanning between the foundation and the roof; part of the wind load is taken directly to the foundation, and the remainder is distributed via the roof plane truss (or *roof bracing*) to the end structural frames. In this way we could construct quite long buildings, with only the end walls being rigid and roof bracing transmitting all the intermediate wind load to these walls.

6.4 ASYMMETRY

Up to this point we have been assuming that the vertical bracing has been arranged symmetrically with respect to the wind load, i.e. in our rigid box analogy, the two end walls share equally in resisting the wind load because they are equidistant from the resultant wind thrust (Figure 6.5). That is, the centre of resistance of the bracing lies on the line of action of the resultant wind force.

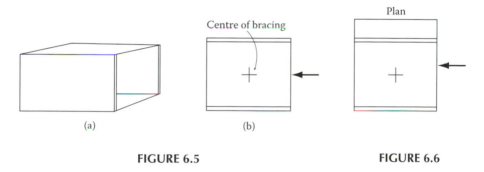

FIGURE 6.5 FIGURE 6.6

On occasion, as in Figure 6.6, the architectural planning of the building means that this cannot be done, and one shear wall or rigid frame or braced frame will take more wind load than another.

In extreme cases, where the centre of resistance of the bracing is quite some distance from the resultant wind force, the building as a whole may be subject to extensive twisting moments, i.e. the bracing system must be able to act both as a vertical cantilever in bending and as a rigid tube to resist torsion (Figure 6.7). Tall buildings, in which an eccentric lift shaft is the major bracing system, can be subjected to torsion in this way.

LOAD-PATHS

In a single storey shed-type building, wind loads produce stresses in many parts of the structure. We often describe this (somewhat loosely) by trying to trace the *path* of the load from the point of application to the foundations. For example, wind load on a sheet of wall cladding may be thought to be transmitted along the following path: vertically into the girts, then horizontally into the columns; vertically along the column, partly into the foundation, the remainder into the roof bracing; horizontally through the roof bracing and then finally vertically down through the end-wall bracing to the foundation.

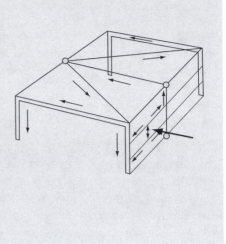

WORKSHEET 6.1

In this chapter we have studied many ways in which buildings may be braced against lateral loads. Diagrams A to V illustrate many of the methods that can be adopted for this purpose. It is important that we understand the practical application of these methods.

Try to find two structures that utilise each of the methods of bracing described in this chapter, and make sure that you understand how each structure behaves when horizontal loads are applied to it. Try to choose simple structures that you can study closely; furniture, bus shelters, service stations and factories may be suitable. Remember that the horizontal load may be applied from *any* direction; often different bracing systems may be used in the one structure.

Record your observations for each pair of structures by using photographs and diagrams to show the paths by which wind load is transmitted to the foundation.

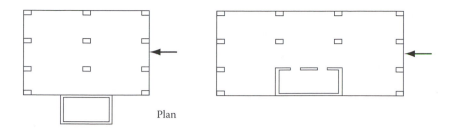

Plan

FIGURE 6.7

6.5 COMBINED EFFECTS

We see then, that many structural members in a building participate in transmitting two separate loading systems (vertical and lateral) to the foundations, and these members must be designed to fulfil this dual role. Obviously some members are used primarily for just one force system, and their role in the other force system is secondary. The column in Figure 6.8, for example, will be used primarily to carry vertical loads, but it may also have bending moments induced by wind loading, and must, of course, be designed for the most severe combination of both loads.

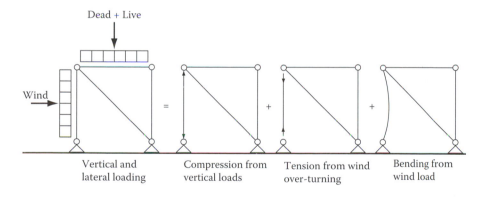

| Vertical and lateral loading | Compression from vertical loads | Tension from wind over-turning | Bending from wind load |

FIGURE 6.8

APPLICATION: BRACING, OSAKA INTERNATIONAL CONFERENCE
CENTRE (SEE PHOTOGRAPHS OPPOSITE)

The Osaka International Conference Centre (OICC) is typical of buildings in highly seismic zones, in requiring a strong, stiff, and ductile stability system to resist earthquake forces. Code requirements in such zones can lead to horizontal design forces more than half of a building's weight. Much structural design effort goes into ways of justifying reduced design forces, through increased structural ductility or damping or base isolation.

The seismic stability system for OICC uses several sets of diagonal braces. These can be made very stiff and strong. To improve their ductility — their ability to absorb seismically induced energy without breaking — 'unbonded' braces were specified. These have an internal steel member 'unbonded' from a surrounding concrete encasement that prevents buckling under high compression, and so allows the internal member to undergo plastic yielding without breaking, in extreme seismic events.

Project: Osaka International Conference Centre, Osaka, Japan; Architect: Kisho Kurokawa Architect and Associates; Structural Engineer: Arup; Photos/diagrams: Arup

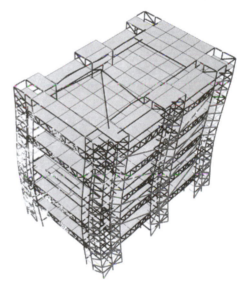

APPLICATION: BRACING, CHINA CENTRAL TELEVISION
(SEE PHOTOGRAPHS OPPOSITE)

To handle the large forces generated by the unusual 'leaning towers' of the China Central Television (CCTV) building in Beijing, an early decision was made to wrap a fully cross-braced steel framed 'tube' around each segment of the building. Although only the diagonal members of these braced tubes are expressed in the facade treatment, structurally they act in unison with 'vertical' columns (typically inclined at 6 degrees, like the towers themselves) and horizontal floor edge beams, via the 'butterfly plates' provided at joints where all these members meet.

The pattern of diagonal bracing was chosen with help from a simple algorithm whereby highly stressed zones in an initially uniform bracing mesh were 'densified' (mesh frequency doubled) and lowly stressed zones were 'rarefied' (mesh frequency halved) until a pattern emerged that was both visually and structurally satisfactory. In this unusual structure, the diagonal bracing responds not just to lateral loads like wind and earthquake, but to gravity loads as well, particularly in providing support to the 'overhang' segments at high level.

Project: China Central Television (CCTV), Beijing; Architect: Office of Metropolitan Architecture (OMA); Structural Engineer: Arup, with ECADI (LDI); Photos: Arup

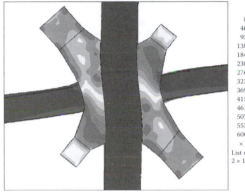

0.00
46.15
92.31
138.46
184.62
230.77
276.92
323.08
369.23
415.38
461.54
507.69
553.85
600.00

× 1.0E+06
List of loadcases
2 × 1.00000 E+00

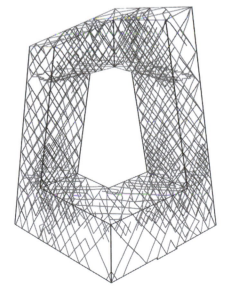

7 Shear Force and Bending Moment

SHEAR FORCE

Shear stress was discussed in Chapter 2. The
action of a pair of scissors on a sheet of paper is
the simplest example of shear that will be familiar
to you. The shearing action of the scissors causes
one part of the paper to slide across the adjacent
part. If the sheet of paper does not have enough
strength to resist the applied shear force, it will
fail, and will be cut by the scissors.

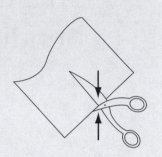

A bolt used to connect metal plates may also
have shear forces applied to it, and will fail if the
forces become too large.

The shear that occurs in a beam is the same as the shear in the sheet of paper
or the bolt; if they seem different it is only because the distance separating the
applied forces is much larger for the beam. All three are covered by the definition in paragraph 7.2.

EXAMPLE 7.1

Draw the Shear Force (S.F.) Diagram for the beam shown:

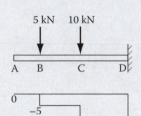

S.F. at A = 0

S.F. just to left of B = 0

S.F. just to right of B   = –5 kN

S.F. just to right of C   = –5 + (–10)

                          = –15 kN

S.F. between C and D = –15 kN

(Note values shown on diagram)

CONTENT OF CHAPTER 7

Although structures composed of axially stressed members, as described in Chapters 2, 4 and 5, are the most efficient class of structure, they are not the most widely used structures in architecture. Most buildings use flexural members, members that function in bending, as the dominant structural elements. Beams, girders, rigid frames and portals are common examples of flexural elements, and in this chapter we commence our study of flexure.

Bending and shear are defined quantitatively, and methods are introduced for calculating and plotting the Bending Moment Diagram (BMD) and Shear Force Diagram (SFD) for a beam.

7.1   TERMS

In Chapter 1 we encountered the terms Shear Force and Bending Moment and saw that they represent the effects produced by external forces acting upon a structure. In this chapter, we will examine these effects in greater detail.

7.2   SHEAR FORCE

If we consider any cross-section A on the axis of a beam, we see that in general there will be a number of forces acting perpendicularly to the axis of the beam to the left of A, and a number acting perpendicularly to the right of A (Figure 7.1).

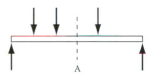

FIGURE 7.1

Definition

> The *Shear Force* at any cross-section A in a straight beam is the algebraic sum of all force components acting perpendicularly to the axis of the beam to *either* side of A.

The *sign* given to a Shear Force depends upon the type of potential relative movement produced by the Shear Force. If the Shear Force at A is such that it tends to move the right-hand portion of the beam downwards with respect to the left-hand portion, the Shear Force is Positive (Figure 7.2).

Positive shear   Negative shear

FIGURE 7.2

EXAMPLE 7.2

Draw the S.F. Diagram for the beam shown.

Replace the distributed load by an equal point load to calculate reactions:

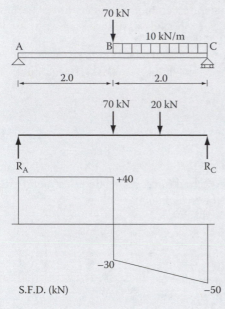

$\sum M_A = 0 = 70 \times 2 + 20 \times 3 - R_C \times 4$

$\therefore R_C = \dfrac{200}{4} = 50 \text{ kN}$

$\sum M_C = 0 = R_A \times 4 - 70 \times 2 - 20 \times 1$

$\therefore R_A = \dfrac{160}{4} = 40 \text{ kN}$

Check $\sum V = 0 = 50 + 40 - 70 - 20$

$\therefore$ OK

S.F. between A and B = +40 kN

S.F. just to right of B   = +40 − 70

  = −30 kN

S.F. just to left of C   = −30 − (10 × 2)

  = −50 kN

Between B and C, the diagram slopes at 10 kN per metre.

BENDING MOMENT

The moment of a force is the product of the force times the lever arm. If you hold a weight in your hand, and then extend your arm straight out from your shoulder, the weight will produce moments along your arm. Because your arm is experiencing bending, we call the moment that is being applied to your arm a Bending Moment (B.M.). The Bending Moment is not constant at all points along your arm, of course; it is very small near your wrists, and very large at your shoulder. The load you are holding in your hand is not the only force that is contributing to the Bending Moment applied to your arm; even if you removed this load, the weight of your arm itself (a distributed load) would still be producing bending in your arm.

If we plot the value of the Shear Force at all points along the axis of a beam, we obtain a Shear Force Diagram. The Shear Force Diagram (S.F.D.) is essentially a graph, with Shear Force (S.F.) plotted along the vertical axis and distance from the end of the beam plotted along the horizontal axis. By considering sections just to the left and just to the right of a point load, we see that at a point load, the S.F changes instantaneously, and the S.F. diagram shows a 'step' (Figure 7.3). By considering a series of sections at equal intervals along a uniformly distributed load, we see that the shear force changes by equal increments, and the Shear Force diagram will have a constant slope (see Examples 7.1 and 7.2).

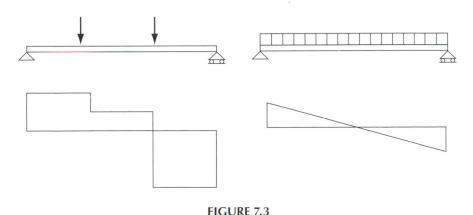

FIGURE 7.3

The Shear Force at any section of a beam provides us with a measure of the tendency for the left-hand part of the beam to slide past the right-hand part. At sections where the Shear Force is large, the tendency for a shear-type movement is large, and we may have to check that the beam is strong enough to resist this tendency. At sections where the S.F. is small, there is little tendency for this relative sliding to occur. The Shear Force Diagram is important because it enables us to see at a glance the parts of the structure that may fail in shear.

7.3   BENDING MOMENT

In paragraph 1.5, we defined the *moment* of a force about a point as the product of the force and the perpendicular distance from the point to the force.

Definition

> The *Bending Moment* at any cross-section A in a straight beam is the algebraic *sum of the moments* about A of all forces *to one side of A*.

Thus, referring to Figure 7.1, each of the three forces to the *left* of the cross-section A produces a moment about A equal to the product of the force times the perpendicular distance from A. The sum of these three moments is called the 'Bending Moment at A'.

BENDING MOMENT   (CONTINUED)

Thus, the B.M. at your shoulder is made up of two components: the moment produced by the weight your hand is holding and the moment produced by the self-weight of your arm. If other loads were applied to your arm, they also would contribute to the B.M. The B.M. at your shoulder is the sum of the moments produced by all forces between your shoulder and your finger tips.

EXAMPLE 7.3

Draw the Bending Moment Diagram for the cantilever shown.

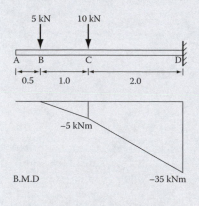

The B.M. at all sections between A and B is zero, because there are no forces to the left of B.

At points to the right of B, the 5 kN force will produce moments as follows:

At 0.2 m from B, B.M. = −5 × 0.2
$$= -1.0 \text{ kNm}$$

At 0.4 m from B, B.M. = −5 × 0.4
$$= -2.0 \text{ kNm}$$

At 0.6 m from B, B.M. = −5 × 0.6
$$= -3.0 \text{ kNm}$$

At 1.0 m from B, B.M. = −5 × 1.0
$$= -5.0 \text{ kNm}$$

(Notice that a concentrated load produced Bending Moments that vary linearly along the beam.)

At 0.5 m to the right of C, B.M. = −5 × 1.5 − 10 × 0.5 = −12.5 kNm
At 1.0 m to the right of C, B.M. = −5 × 2.0 − 10 × 1.0 = −20.0 kNm
At 1.5 m to the right of C, B.M. = −5 × 2.5 − 10 × 1.5 = −27.5 kNm
At 2.0 m to the right of C, B.M. = −5 × 3.0 − 10 × 2.0 = −35.0 kNm

Alternatively, we could have considered the two forces to the *right* of A; the sum of the moments about A of these two forces is also, by definition, the Bending Moment at A, and, of course, will have the same value.

The *sign* given to the Bending Moment produced by a force depends upon the curvature produced by the force. A force that tends to curve the beam so it is concave upwards is said to produce a *positive* bending moment (Figure 7.4). From this we draw the important inference that *upward forces produce* positive Bending Moments.

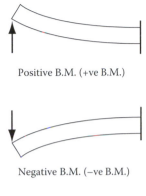

Positive B.M. (+ve B.M.)

Negative B.M. (−ve B.M.)

FIGURE 7.4

The Bending Moment at a section in a structure is important because it gives us a measure of the extent to which the beam is being bent at that section. A large Bending Moment will produce more bending (i.e. sharper curvatures) than would a small Bending Moment. Failure in bending would be expected to occur at parts of the beam where the Bending Moments are large.

In fact, it should be noted that, in Structures, we are not really interested in the forces acting on a beam *per se*; we are much more interested in the *effects* that these forces produce. The concepts of S.F. and B.M. are so useful because they measure these effects; they measure the *shearing* effect (i.e. the tendency of one part of a beam to be translated with respect to the other part) and the *bending* effect (i.e. the tendency for one part to be rotated with respect to the other part). The calculation of S.F. and B.M. is made just so that we may replace a complicated assortment of miscellaneous external forces and loads by the two simple internal effects that they produce.

The maximum value of the B.M. in a beam, whether positive or negative, is then the maximum bending effect in the beam. The point at which it occurs is therefore the point at which the beam will be most highly stressed in bending; it is the point where the beam will break if the applied forces become too large. For this reason the maximum B.M. is an extremely important characteristic of a loaded structure. The value of the maximum, and its location, should always be computed.

On the other hand, for some loadings, a beam may have a location for which the B.M. is zero (Figure 7.5). At this point there will be no bending stresses in the beam. Usually, to one side of this point the beam will be concave downwards with tension stresses at the top surface; to the other side of this point the beam will be concave upwards with compressive stresses at

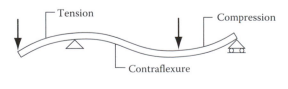

FIGURE 7.5

EXAMPLE 7.4

Draw the S.F.D. and the B.M.D. for the beam shown.

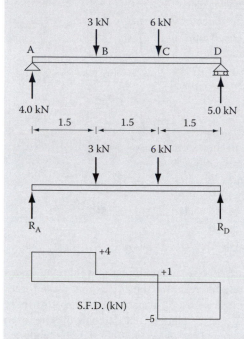

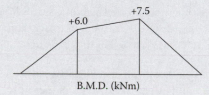

S.F.D. (kN)

B.M.D. (kNm)

1. Draw a simplified diagram showing the reactions at the supports (R_A and R_D)

2. Compute the reactions:

$$\Sigma M_A = 0 = 3 \times 1.5 + 6 \times 3.0$$
$$- R_D \times 4.5$$
$$\therefore R_D = \frac{4.5 + 18.0}{4.5} = 5.0 \text{ kN}$$
$$\Sigma M_D = 0 = R_A \times 4.5 - 3 \times 3.0$$
$$- 6 \times 1.5$$
$$\therefore R_A = \frac{9.0 \times 9.0}{4.5} = 4.0 \text{ kN}$$

Check $\Sigma F_y = 0 = 5.0 + 4.0$
$$- 3.0 - 6.0$$

OK

3. Show these reactions on the sketch of the beam, and draw the S.F.D.

4. Compute Bending Moments:

$$BM_A = 0$$
$$BM_B = 4.0 \times 1.5$$
$$= +6.0 \text{ kNm}$$
$$BM_C = 4.0 \times 3.0 - 3.0 \times 1.5$$
$$= +7.5 \text{ kNm}$$
$$OR = 5.0 \times 1.5 = 7.5 \text{ kNm}$$
$$BM_D = 4.0 \times 4.5 - 3.0 \times 3.0$$
$$- 6.0 \times 1.5$$
$$= 0$$

5. Draw the B.M.D.

the top surface. Such a point is called a *point of contraflexure*; the curvature of the beam changes sign, and the B.M.s and stresses also change sign.

7.4 BENDING MOMENT DIAGRAMS

If we plot the value of the Bending Moment at all points along the axis of a beam, we obtain a Bending Moment Diagram. This is a graph in which B.M. is plotted vertically and distance along the beam is plotted horizontally. We shall adopt a convention that *positive Bending Moments will be plotted above the x-axis*. (This is a common, though by no means uniform, convention.)

Example 7.3 shows how a B.M.D. can be plotted by considering a series of points along the beam. The important inference that can be drawn from this example is that the B.M.D. for *concentrated loads* consists of straight lines. To draw the B.M.D., we need to compute only the B.M.s at the points of application of the forces, and these plotted points may then be joined by straight lines. This procedure is illustrated in Example 7.4. The format of this example should be particularly noted. The various diagrams should be below one another, and to one side of the page, so that the computations can be developed along with the diagrams.

The Bending Moment Diagram for a beam carrying *distributed loads* is obtained in just the same way as for concentrated loads. We consider a series of sections along the beam, and compute the B.M. at each section for the loads to one side only. We replace *that part* of a distributed load to one side of the section by an equivalent point load. Example 7.5 illustrates the method. We see from this example that the B.M.D. produced by a uniformly distributed load is a parabola. If the uniformly distributed load is applied to only part of the length of the beam, the B.M.D. will be part of a parabola. In order to sketch the B.M.D. for a beam that supports a distributed load, it is usually sufficient to compute the B.M. at only two or three points, one at each of the loads and perhaps one near the middle (see Example 7.6). Consequently, we see that the B.M.D. for uniformly distributed loads consists of curved lines; for concentrated loads, the diagram consists of straight lines. When both types of loads occur in a single beam, the diagram may have curved and straight lines. Usually the diagram can be drawn from values computed at certain critical locations only.

7.5 LOAD, SHEAR FORCE AND BENDING MOMENT

A mathematical relationship exists between Load, S.F. and B.M.:

Consider a beam AB (Figure 7.6) carrying a uniformly distributed load of intensity w N/m over part of its length. Consider a point C distant x from A, and assume that

EXAMPLE 7.5

Deduce the equation for the Bending Moment Diagram for a simply-supported beam carrying a uniformly distributed load of w N/m over a span of L m.

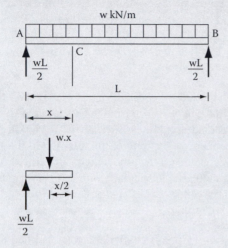

By symmetry:

$$R_A = R_B = \frac{wL}{2}$$

At any point C, x m from A, the bending moment is given by:

$$BM_x = \frac{wL}{2}.x - \frac{wx^2}{2}$$

This is the equation of a second degree parabola, i.e., the B.M. Diagram will be a parabola, with a maximum value (at $x = \frac{L}{2}$) of $\frac{wL^2}{8}$.

Hence, for a simply supported beam carrying U.D. loading:

$$BM_{max} = \frac{wL^2}{8} = \frac{WL}{8} \quad \text{where} \quad w = \text{load per m}$$

$$W = \text{total load}$$

at C the S.F. is S and the B.M. is M. Consider a second point D, a very small distance dx from C, and assume that at D the S.F. has changed to S + dS, and the B.M. has changed to M + dM.

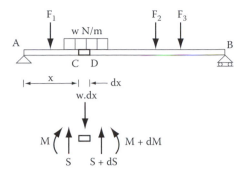

FIGURE 7.6

Now, the *actual change* in S.F. between C and D is the intensity of load multiplied by the distance, i.e. *the S.F. at D* is S + w.dx. Hence:

$$S + dS = S + w.dx \quad \text{or} \quad \frac{dS}{dx} = w \quad (1)$$

Likewise, the *actual change* in B.M. between C and D is the S.F. at C times the distance dx, plus the moment D of the distributed load w, i.e. *the B.M. at D* is $M + S.dx - w.dx.\dfrac{dx}{2}$.

Hence $M + dM = M + S.dx - w.dx.\dfrac{dx}{2}$.

Neglecting second order differentials, we have $\dfrac{dM}{dx} = S$ (2)

Equations 1 and 2 are extremely helpful in drawing S.F. and B.M. diagrams (Figure 7.7).

Equation 1 states that the rate of change of S.F. at a point is equal to the intensity of load at that point, i.e. *the slope of the S.F. Diagram equals the intensity of the load.* If, over a part of a beam, there is no load, the slope of the S.F. Diagram will be zero, i.e. the diagram is parallel to the beam axis. If, at a point in a beam, there is a point load, then the intensity of loading is infinitely large and the slope of the S.F. Diagram will be infinite, i.e. the diagram is perpendicular to the beam axis. If, over part of a beam, the load is uniformly distributed, the slope of the S.F. Diagram is uniform.

Equation 2 states that the rate of change of B.M. at a point is equal to the value of the S.F. at that point, i.e. *the slope of the B.M. diagram equals the value of the S.F.* If, over part of a beam, the S.F. is constant, the slope of the B.M. Diagram is constant. If, at a point on the beam, the S.F. is zero, the slope of the B.M. Diagram is zero, i.e. the B.M. reaches a maximum value. (Note this especially: *The B.M. attains a peak value when the S.F. is zero.*)

EXAMPLE 7.6

Draw the S.F.D. and the B.M.D. for the beam shown.

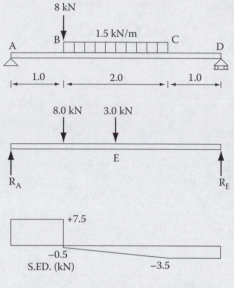

1. Draw a simplified diagram showing the reactions at the supports and also the concentrated load equivalents of the distributed loads.

2. Compute the reactions:

$$R_A = \frac{8 \times 3 + 3 \times 2}{4} = 7.5 \text{ kN}$$

$$R_D = \frac{8 \times 1 + 3 \times 2}{4} = 3.5 \text{ kN}$$

Check $\Sigma F_y = 0$ OK

3. Show these reactions on the sketch of the beam and draw the S.F.D.

4. Compute Bending Moments:

$$BM_B = 7.5 \times 1.0 = 7.5 \text{ kNm}$$

$$BM_C = +7.5 \times 3.0 - 1.5 \times 2$$
$$\times 1.0 - 8.0 \times 2$$
$$= 22.5 - 16 - 3.0$$
$$= 3.5 \text{ kNm}$$

$$BM_E = 7.5 \times 2 - 8 \times 1 - 1.5$$
$$\times 1.0 \times 0.5$$
$$= +6.25 \text{ kNm}$$

5. Draw the B.M.D.

SUMMARY

The importance of the concepts of Bending Moment and Shear Force in architectural structures cannot be overemphasised. Likewise, the ability to sketch quickly the shapes of B.M. and S.F. diagrams is a skill that you should cultivate. These diagrams show the *effects* of the loads on the structure, and are of great importance to the designer.

From a consideration of the discussion in this chapter, we may draw a number of conclusions that will assist us in drawing B.M. and S.F. diagrams. These conclusions are presented here as a set of principles or hints that should be kept in mind when solving problems:

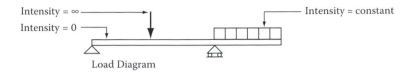

Load Diagram

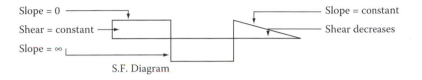

S.F. Diagram

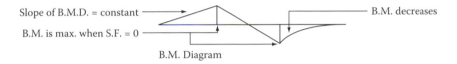

B.M. Diagram

FIGURE 7.7

If we consider two points C and D on a beam, and then integrate Equation 2 between the limits x = c and x = d, we have:

$$M_D - M_C = \int_c^d S \, dx$$

i.e. the change in B.M. between any two points is equal to the integral of shear between those two points, i.e. *the change in B.M. between any two points is equal to the area under the S.F. Diagram between those two points.* Two important applications of this principle are as follows (Figure 7.8):

(a) If the B.M. is zero at the ends of a simply supported beam (as is usually the case), the total area of the S.F. Diagram for the whole beam must be zero, i.e. the area of the negative part of the S.F. Diagram equals the area of the positive part.

(b) If the B.M. is zero at one end of a beam, then the distance from that end to any point of contraflexure can likewise be found by equating the positive and negative shear diagram areas.

SUMMARY (CONTINUED)

1. Always draw all diagrams vertically below each other on the one sheet. Unless specifically requested otherwise, draw diagrams freehand, approximately to scale.

2. Commence with a 'picture' diagram of the beam showing all known loads. Below this draw a simplified load diagram replacing distributed loads by their equivalent point loads. Calculate the reactions from this simplified diagram, *and show the reactions on the first 'picture' diagram. Do not refer to the simplified diagram again.*

3. From the 'picture' diagram draw the S.F. Diagram by commencing at the left-hand end, and plot upwards for upward forces and downwards for downward forces.

4. Locate positions of zero S.F., as these will also be positions of maximum B.M. (slope of B.M. = value of S.F.).

5. Calculate the value of the B.M. at the following significant positions:

   (a) at every support
   (b) at every point load
   (c) at the start and end of every distributed load
   (d) at points of zero shear force.

6. Between these significant positions, draw suitable curves in accordance with the following:

   (a) if the beam is unloaded, the B.M. Diagram varies linearly;
   (b) if the beam carries uniform loading, the B.M. Diagram varies parabolically.

7. Remember:

   (a) upward forces produce positive B.M.s; downward forces produce negative B.M.s;
   (b) a point load produces abrupt changes in S.F., but merely slope changes in the B.M. Diagram;
   (c) in calculating the B.M. at a particular section, any relevant portion of a uniformly distributed load may be replaced by an equal concentrated load acting at the centre of gravity of the portion of the distributed load;
   (d) Bending Moments are always zero at the ends of a beam (unless an applied moment acts at the end).

8. Always sketch the deflected shape that would result from the diagrams you have drawn.

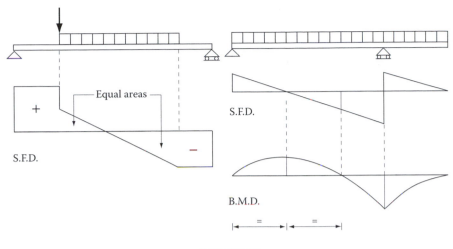

FIGURE 7.8

7.6 CONTRAFLEXURE

The location of any point of contraflexure in a beam is important. Because the bending stresses are zero at a point of contraflexure, we often choose this point to join short lengths together to make a long beam. The joint is easier to make because the bending stresses are small.

Three methods are commonly used to calculate the position of a point of contraflexure:

1. If the loading is uniformly distributed, it is often possible to use the area of the S.F.D., as in the previous paragraph (Figure 7.8).
2. For beams supporting only concentrated loads, it is usually easiest to calculate the location of the point of contraflexure by using the properties of similar triangles. For example, in Figure 7.9a, the distance x is obtained from the two similar triangles as follows:

$$\frac{x}{2.0} = \frac{60}{140}$$

3. In the third method, the critical bending moments are first computed, and the B.M.D. is drawn (Figure 7.9b). Let x be the distance from the end of the beam to the point of contraflexure. Set up an equation in x for Bending Moment near the point of contraflexure, and solve this equation for the particular case of B.M. = 0. If distributed loads are involved, the equation will be a quadratic of the form:

$$a.x^2 + b.x + c = 0$$

WORKSHEET 7.1

7.1 For each of the beams shown in Figures 1 to 4, calculate for and draw the following diagrams below each other on a single page:

(a) the loadings on the beam
(b) the Shear Force Diagram to a suitable scale
(c) the Bending Moment Diagram to a suitable scale
(d) the anticipated deflected shape of the beam to an exaggerated scale.

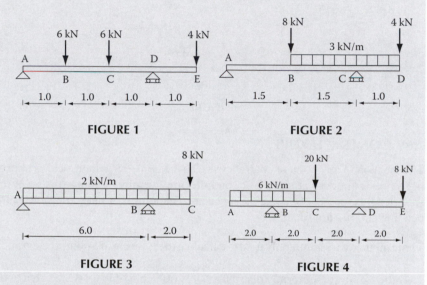

FIGURE 1   FIGURE 2

FIGURE 3   FIGURE 4

7.2 Compute the locations of points of contraflexure in Figure 4.

7.3 In Figure 5 are sketched a number of beams and the loads applied to them. For each of these, without performing any calculations, sketch the shape of the S.F.D. and B.M.D. that you would anticipate. (Note: Because the magnitudes of the loads are not given, there can be no uniquely correct answers to this question.)

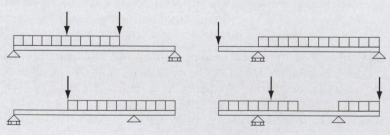

FIGURE 5

for which the solutions are:

$$x = \frac{-b \pm \sqrt{b^2 - 4\,ac}}{2a}$$

Only one of these two solutions will have meaning in terms of the problem. This third method is quite general in its application, and is not restricted to particular loading conditions (as are the first two methods).

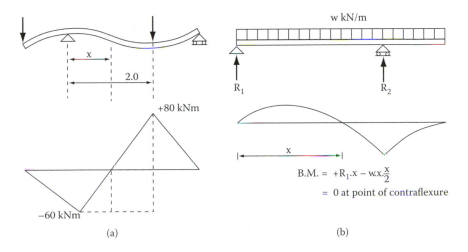

FIGURE 7.9

APPLICATION: SFD, BMD, RED CENTRE (SEE PHOTOGRAPHS OVER PAGE)

The West Wing of the Red Centre at the University of New South Wales uses deep transfer beams ABC at 2nd floor level to collect point loads D and E from terminating columns supporting floor slabs at floors 3, 4, and 5. This creates a grand entry portal in the north wall, free of facade columns.

For preliminary design, the twin columns at B could be represented by a single reaction, with another vertical reaction at C. The load from column F is assumed to pass directly into the column beneath. The point loads at D and E then give rise to a Shear Force Diagram and a Bending Moment Diagram like those indicated. There is also a uniformly distributed load (UDL) on AC from the beam self-weight, and from the floor slab adjacent to the beam, both considerably smaller than the point loads. What effect would the UDL have on the S.F.D. and B.M.D.?

Project: Red Centre, University of New South Wales, Sydney, Australia; Architect: MGT Architects; Structural Engineer: Arup; Photos/diagrams: Arup

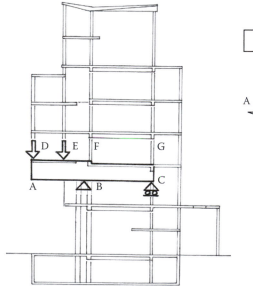

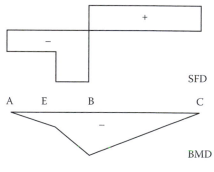

SFD

A   E   B   C

BMD

8 Properties of Area

CENTRE OF GRAVITY

As stated in paragraph 8.1, the Centre of Gravity is the point where the entire weight of the object *appears* to be concentrated. This is the principle that underlies the method of computation in Example 8.1. It also provides a convenient experimental method for finding the C.G. If the body is suspended from any point, it will rotate until the C.G. is vertically beneath the point of support. (Can you explain why this is so?) By suspending the body several times from different points, and drawing the vertical through each point of support, the C.G. may be found.

Reaction at support

Weight

EXAMPLE 8.1

The metal plate shown is 10 mm thick and has a density of 8000 kg/m³. What is the location of the centre of gravity?

Weight of section 1

$= 0.080 \times 0.060 \times 0.010 \times 8 \times 10^3$
$\quad \times 10$ (N/kg)
$= 3.84$ N

Weight of section 2

$= 0.120 \times 0.020 \times 0.010 \times 8 \times 10^3$
$\quad \times 10$ (N/kg)
$= 1.92$ N

Weight of section 3

$= 0.060 \times 0.120 \times 0.010 \times 8 \times 10^3$
$\quad \times 10$ (N/kg)
$= 5.76$ N

$\therefore$ Resultant $= 3.84 + 1.92 + 5.76$
$\qquad = 11.52$ N

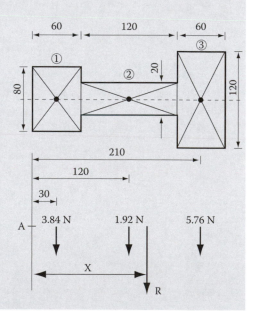

CONTENT OF CHAPTER 8

The behaviour of many structural components depends to a large extent upon their cross-sectional shapes. The amount of deflection of a beam under load, the load that a long column can safely carry, the stability of a retaining wall against overturning — these are just some of the important consequences of the shape of structural cross-sections. In this chapter, we will bring together these attributes of shape and area, so that they can be related to one another.

Concepts introduced include centre of gravity, centroid, second moment of area, and section modulus.

8.1 CENTRE OF GRAVITY

The *centre of gravity* of a body is the point in or near the body through which the resultant attraction of the earth acts for all orientations of the body. It is the point where the entire weight of the object appears to be concentrated.

To locate the centre of gravity of a body, we may imagine the body to be subdivided into a large number of small particles. The weight-forces of all of these particles then constitute a system of parallel forces, and the location of the resultant can be determined. By turning the body into other orientations, we can determine the line of action of the resultant in each case. The C.G. lies at the point of intersection of these lines.

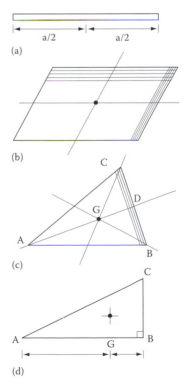

Some particular instances are:

(a) A thin rod of length 'a' (Figure 8.1a): The C.G. is halfway along the length.
(b) A thin parallelogram-shaped plate (Figure 8.1b): By considering the plate to be composed of a large number of thin rods parallel to each side in turn, we see that the C.G. is at the intersection of the bisectors of the sides, which coincides with the *intersection of the diagonals.*
(c) A thin triangular plate (Figure 8.1c): By considering the plate to be composed of a large number of thin rods parallel to each side in turn, we see that the C.G. is at the *intersection of the medians.* It can be proved that

FIGURE 8.1

EXAMPLE 8.1   (CONTINUED)

Take moments about A:

$$\therefore R \times X = 3.84 \times 30 + 1.92 \times 120 + 5.76 \times 210$$

$$\therefore X = \frac{1555}{11.52} = 135$$

i.e. the C.G. is on the horizontal axis of symmetry, 135 mm from L.H. end.

EXAMPLE 8.2

A wall with the cross-section shown has a density of 2400 kg/m³. The soil behind the wall applies a horizontal force of 80 kN per metre length of wall. Will the wall overturn?

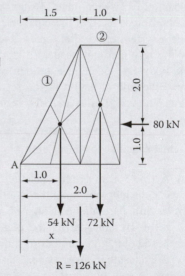

Per metre length of wall, we have:

Weight of (1) $= \dfrac{3 \times 1.5}{2} \times 1 \times 2400 \times 10 = 54$ kN

Weight of (2) $= 3 \times 1 \times 1 \times 2400 \times 10 = 72$ kN

$\therefore$ Resultant weight $= 126$ kN

Taking moments about A:

$$x = \frac{54 \times 1 + 72 \times 2}{126} = 1.57\,\text{m}$$

i.e. the C.G. is 1.57 m from point A. If the wall were to rotate about point A, we would have:

Restraining Moment about A

$\qquad = 126 \times 1.57 = 198$ kNm

Overturning Moment about A

$\qquad = 80 \times 1 = 80$ kNm

$\therefore$ The wall will not overturn about A.

$$DG = \frac{DA}{3}$$

For a right-angled triangular plate:

$$BG = \frac{BA}{3}$$

Since the resultant gravitational force passes through the C.G. of a body, we may find the C.G. of a complex body by dividing it into a number of simple bodies (such as those considered above) and determining the location of the C.G. of each. The separate gravity forces may then be replaced by a single resultant force and the location of the C.G. for the complex body determined by taking moments about any point (refer paragraph 1.12). If the body has no axes of symmetry, this procedure will need to be repeated for other orientations of the body. Examples 8.1 and 8.2 illustrate the method.

8.2 THE CENTROID

It will be noted that the calculations of Example 8.1 would not have been altered if the body had been 5 mm or 20 mm or 100 mm thick; the calculations we performed were essentially on the *plan area* of the plate. In fact, it is often convenient to determine the 'centre of gravity' of a weightless area. We take moments of small areas about a point instead of moments of forces. The 'centre of gravity' of an area is called the *centroid*.

A recommended procedure to be used in determining the centroid of an area is as follows:

(a) Select a pair of rectangular axes which pass through the extremities of the area.
(b) Sub-divide the area into simple geometrical shapes. Determine the area 'a' of each shape, and the distance y of its centroid from the x axis.
(c) Calculate the moment 'a.y' of each shape about the x axis.
(d) The distance $\bar{y}$ from the axis to the centroid of the whole area is then found by summing the moments and dividing by the total area:

$$\bar{y} = \frac{\Sigma a.y}{\Sigma a}$$

(e) Repeat for the y axis, i.e. $\bar{x} = \frac{\Sigma a.x}{\Sigma a}$.

The operation can frequently be performed more conveniently in tabular form (Example 8.3).

EXAMPLE 8.3

Determine the location of the centroid of the plane area shown.

By symmetry, $\overline{X} = 230$ mm

$$\overline{y} = \frac{\Sigma ay}{\Sigma a}$$

$\Sigma ay = 80 \times 160 \times 80 + 60 \times 300$
$\qquad \times 30 + 80 \times 160 \times 80$
$\qquad = 2588 \times 10^3$

$\Sigma a = 80 \times 160 + 60 \times 300$
$\qquad + 80 \times 160$
$\qquad = 43.6 \times 10^3$

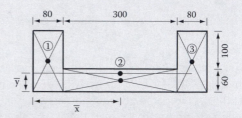

Or, in tabular form:

Area	a	y	ay	
(1)	$80 \times 160 = 12\ 800$	80	1 024 000	$\overline{y} = \dfrac{2588 \times 10^3}{43.6 \times 10^3}$
(2)	$60 \times 300 = 18\ 000$	30	540 000	
(3)	$80 \times 160 = 12\ 800$	80	1 024 000	
Σ	43.6×10^3	–	2588×10^3	$= 59.4$ mm

EXAMPLE 8.4

Calculate the location of the centroid, and the value of I_{xx} for the shape shown.

To calculate the location of the centroid, choose a reference axis passing through the lower extremity of the shape (axis x_1x_1). Subdivide into two areas 1 and 2, and set up the following table:

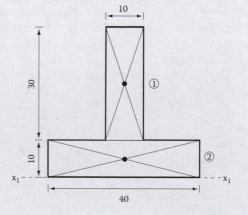

Area	a	y	ay
1	300	25	7500
2	400	5	2000
Σ	700		9500

The term Σay is sometimes called the *First Moment* of the area A about the reference axis xx. If the particular xx axis chosen happens to pass through the centroid (i.e. is a *centroidal axis*) Σay = 0, because some of the y-distances will be negative.

8.3   SECOND MOMENT OF AREA

The Second Moment of Area (or the Moment of Inertia as it is sometimes called) is a property of area of great importance in Mechanics.

If we imagine an area A (Figure 8.2) to be sub-divided into a large number of infinitesimal areas 'a', the Second Moment of Area of A about the axis xx is defined as Σay^2 (i.e. the sum of the products formed by multiplying each small area 'a' by the square of its distance from the xx axis). The Second Moment is always positive (because it contains the term y^2). It is given the symbol I and has the units (length)⁴, e.g. mm⁴.

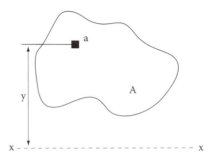

FIGURE 8.2

$$I = \Sigma ay^2$$

The Second Moment of Area may be calculated with respect to *any* axis, but in Structures we use it specifically for the horizontal and vertical centroidal axes (i.e. the axes that pass through the centroid). Thus we will use the notation I_{xx} to refer to the Second Moment of an area about its horizontal centroidal axis.

The Second Moment may be calculated for any mathematically definable area by calculus. The rectangle is a particularly important example, and the value of the Second Moment for the rectangle shown in Figure 8.3 is (see page 122):

$$I_{xx(rect)} = \frac{bd^3}{12}$$

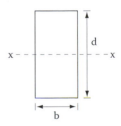

FIGURE 8.3

This result is particularly important and should be committed to memory. Thus, for a rectangle 30 mm wide and 40 mm deep, $I_{xx} = 30 \times 40^3/12 = 0.160 \times 10^6$ mm⁴.

For other simple shapes, similar expressions may be derived, and some are listed in Table 11.1.

EXAMPLE 8.4   (CONTINUED)

$$\bar{y} = \frac{\Sigma ay}{\Sigma a}$$

$$= \frac{9500}{700} = 13.6 \text{ mm}$$

Hence the centroid is on the vertical axis of symmetry, 13.6 mm from the base.

To calculate I_{xx} where xx is the horizontal axis through the centroid, use the Theorem of Parallel Axes for each of the two simple rectangles. (Why don't we use the Common Centroid Theorem?)

$$I_{xx} = I_{zz} + Ae^2$$

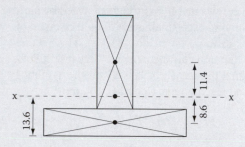

For rectangle 1,

$$I_{xx1} = \frac{10 \times 30^3}{12} + 10 \times 30 \times 11.4^2$$

$$= 22.5 \times 10^3 + 39.0 \times 10^3$$

$$= 61.5 \times 10^3 \text{ mm}^4$$

For rectangle 2,

$$I_{xx2} = \frac{40 \times 10^3}{12} + 40 \times 10 \times 8.6^2$$

$$= 3.3 \times 10^3 + 29.6 \times 10^3$$

$$= 32.9 \times 10^3 \text{ mm}^4$$

$$\therefore \ I_{xx} = I_{xx1} + I_{xx2}$$

$$= (61.5 + 32.9) \times 10^3 \text{ mm}^4$$

$$= 94.4 \times 10^3 \text{ mm}^4$$

For more complex shapes, it is usual to subdivide the total area into a series of simple shapes, for each of which the value of the Second Moment is known. The following two Theorems may then be used to compute the Second Moment of the composite.

8.4 THEOREM OF PARALLEL AXES

The Second Moment of a simple area A about an axis xx *not* through its centroid equals the Second Moment about a parallel centroidal axis, plus A times the square of the distance between the axes (Figure 8.4). The proof is given on page 124.

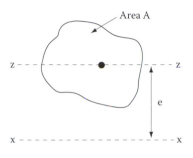

$$I_{xx} = I_{zz} + Ae^2$$

FIGURE 8.4

8.5 COMMON CENTROID THEOREM

When two or more simple areas have *a common centroidal axis*, the Second Moment of the composite shape is the algebraic sum of the Second Moments of the simple areas (Figure 8.5).

The Theorem of Parallel Axes can be used to compute I for *any* area provided that it can be divided into simple areas such as those in Example 8.4, which illustrates the method. The shape is divided into a set of simple areas, and the centroid is found. For each simple area, we find the Second Moment about its own centroidal axis (I_{zz}), and the product of its area (A) times the square of the distance between its own axis and the centroidal axis of the entire shape ($A.e^2$). The Second Moment of the whole shape is then the sum of all such terms.

The Common Centroid Theorem is much simpler to use than the Parallel Axes Theorem, but its use must be restricted to shapes where the simple areas *all have their centroids on the same axis*. In Figure 8.5, for example, this theorem *could not* be used to compute I_{yy}, because the centroids do *not* lie on the same yy axis. Example 8.5 demonstrates the method.

The Second Moment of Area becomes numerically very large for shapes in which much of the area is quite some distance from the centroidal axis. For a rectangle, for example, the Second Moment about the horizontal axis is proportional to the cube of the height.

ALTERNATIVE SOLUTION TO EXAMPLE 8.4

Once the procedure involved in using the Theorem of Parallel Axes is fully understood, the labour in computation can be reduced by adding a column for $e \, (= y - \bar{y})$ in the tabulation. A second column can then be added for $a.e^2$.

The full solution would then appear as follows:

Area	a	y	ay	$e = y - \bar{y}$	ae^2
1	300	25	7500	11.4	39×10^3
2	400	5	2000	-8.6	29.6×10^3
Σ	700		9500		

$$\bar{y} = \frac{\Sigma ay}{\Sigma a} = \frac{9500}{700} = 13.6 \text{ mm}$$

$$I_{xx} = I_{xx1} + I_{xx2}$$

$$= \frac{10 \times 30^3}{12} + 39 \times 10^3 + \frac{40 \times 10^3}{12} + 29.6 \times 10^3$$

$$= 94.4 \times 10^3 \text{ mm}^4$$

SECOND MOMENT OF AREA OF A RECTANGLE

Consider a thin strip b long and dy wide, a distance y from the axis xx.

The Second Moment of this strip about xx is:

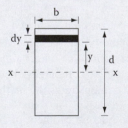

$$a.y^2 = b.dy.y^2$$

$$\therefore \Sigma a.y^2 = 2 \int_{y=0}^{y=d/2} b.y^2.dy = 2[b.y^3/3]_0^{d/2} = \frac{b.d^3}{12}$$

i.e. $I_{xx(rect)} = \dfrac{b.d^3}{12}$

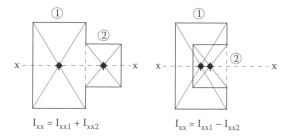

$$I_{xx} = I_{xx1} + I_{xx2} \qquad\qquad I_{xx} = I_{xx1} - I_{xx2}$$

FIGURE 8.5

The major importance of the Second Moment in structural theory is the fact that it provides us with a direct measure of the stiffness of flexural members. The stiffness of a beam is directly proportional to the Second Moment of its cross-sectional shape. Because most building structures need to be stiff, we find that the most popular cross-sectional shapes for beams are those having relatively large Second Moments (for stiffness), as well as fairly small cross-sectional areas (for economy of material use). In Figure 8.6 are shown a series of beam cross-sections, all having the same area, but having markedly different Second Moments.

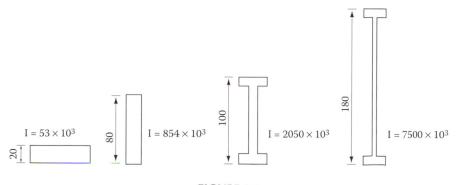

FIGURE 8.6

8.6   SECTION MODULUS

The Section Modulus is yet another property of area that is of great importance in flexural structural members.

The Section Modulus of a shape about a centroidal axis is the Second Moment (about that axis) divided by the distance from the axis to the furthermost part of the shape.

$$\text{Section Modulus } Z_{xx} = \frac{I_{xx}}{c}$$

where c is the distance from axis xx to the top or the bottom of the shape. The units of Z are typically mm^3.

THEOREM OF PARALLEL AXES

The proof is straightforward:

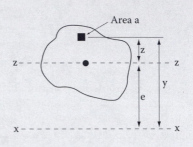

$$I_{xx} = \Sigma a.y^2 = \Sigma a(e + z)^2$$
$$= \Sigma a.e^2 + \Sigma 2a.e.z + \Sigma a.z^2$$
$$= A.e^2 + 2e\Sigma a.z + I_{zz}$$

But, because zz is a centroidal axis:

$$\Sigma a.z = 0 \quad \text{(see paragraph 8.2)}$$
$$I_{xx} = I_{zz} + A.e^2$$

EXAMPLE 8.5

Calculate the Second Moments of Area about the axes xx and yy for the hollow rectangle shown.

By applying the Common Centroid Theorem, we have:

$$I_{xx} = \frac{BD^3}{12} - \frac{bd^3}{12} = \frac{80 \times 120^3}{12} - \frac{60 \times 100^3}{12}$$
$$= (11.52 - 5.0) \times 10^6$$
$$= 6.52 \times 10^6 \text{ mm}^4$$

$$I_{yy} = \frac{120 \times 80^3}{12} - \frac{100 \times 60^3}{12}$$
$$= (5.12 - 1.80) \times 10^6$$
$$= 3.32 \times 10^6 \text{ mm}^4$$

For a rectangle 'b' wide and 'd' deep:

$$I_{xx} = \frac{bd^3}{12} \quad c = \frac{d}{2} \quad \text{and} \quad Z_{xx} = \frac{I_{xx}}{c} = \frac{bd^2}{6}$$

i.e. $Z_{xx(rect)} = \frac{bd^2}{6}$

Where the centroidal axis is also an axis of symmetry, the value of c will be the same for measurements on both sides of the axis. If, however, the shape is *not* symmetrical about the centroidal axis, there will be two possible values of c, and two values of Z (Figure 8.7).

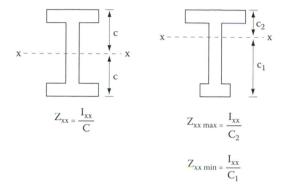

FIGURE 8.7

The Section Modulus, as we shall see in Chapter 9, is of importance in the design of beams. We have pointed out that the Second Moment of the cross-sectional shape of a beam is a direct measure of the *stiffness* of the beam. The Section Modulus of the cross-sectional shape gives a measure of the *strength* of the beam. A beam with a large Section Modulus is stronger, and will support larger loads, than a beam composed of the same material but having a smaller Section Modulus.

EXAMPLE 8.6

Calculate the Section Modulus about the xx axis for the shapes shown in Examples 8.5 and 8.4

$$\text{In Example 8.5} \quad I_{xx} = 6.52 \times 10^6 \text{ mm}^4$$
$$\text{and} \quad c = 60 \text{ mm}$$
$$\therefore \ Z_{xx} = \frac{6.52 \times 10^6}{60} = 109 \times 10^3 \text{ mm}^3$$
$$\text{In Example 8.4,} \quad I_{xx} = 94.4 \times 10^3 \text{ mm}^4.$$

Because the shape is not symmetrical about the xx axis, there are two values of c, viz.,

$$c = 13.6 \text{ mm or } 26.4 \text{ mm}$$
$$\therefore \ Z_{xx} \text{ max} = \frac{94.4 \times 10^3}{13.6} = 6.9 \times 10^3 \text{ mm}^3$$
$$\text{and} \quad Z_{xx} \text{ min} = \frac{94.4 \times 10^3}{26.4} = 3.58 \times 10^3 \text{ mm}^3$$

WORKSHEET 8.1

8.1 The pier shown in elevation in Figure 1 is built of stone having a density of 2000 kg/m³, and supports an arched roof, from which it receives a thrust of 30 kN as shown.

    (a) Determine the location of the centre of gravity of the pier;
    (b) If the pier has a constant thickness of 1 m, determine whether it will be overturned by the roof thrust.

8.2 A beam is to be manufactured from two pieces of timber so as to have the cross-section shown in Figure 2.

    (a) Determine the location of the centroid of the cross-section;
    (b) Determine the values of Second Moment of Area and Section Modulus about *both* vertical and horizontal centroidal axes.

8.3 Figure 3 shows the cross-section of a prestressed concrete girder.

    (a) Determine the location of the centroid;
    (b) Determine the values of I and Z about the horizontal axis.

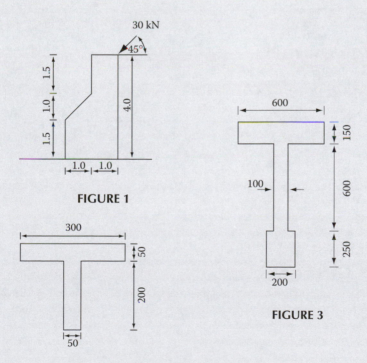

FIGURE 1

FIGURE 2

FIGURE 3

APPLICATION: PROPERTIES OF CROSS SECTIONS, SYDNEY
OPERA HOUSE (SEE PHOTOGRAPHS OPPOSITE)

The ribs that make up the shells of the Sydney Opera House roof fall on great circles of a sphere, so are identical in cross-section at a given distance from a common pole. This allowed all ribs to be cast from a single formwork geometry. The cross-section of each rib increases in width and in depth, towards the ridge. At the base it is almost rectangular, and at the ridge, an open 'Y' shaped section. The cross-section area, second moment of area, and section moduli, vary continuously around the curve of the rib.

While the completed roof behaves like a shell, distributing applied loads by surface action, significant bending moments also appear on the rib cross-sections. These, plus axial forces from applied loads and from the original prestressing, give rise to stress distributions in the ribs that were calculated using the section properties described in this chapter.

Project: Sydney Opera House, Sydney, Australia; Architect: Jorn Utzon; Structural Engineer: Arup; Photos/diagrams: (top): Arup; (bottom left): Yuzo Mikami; (bottom middle): Arup; (bottom right): Yuzo Mikami

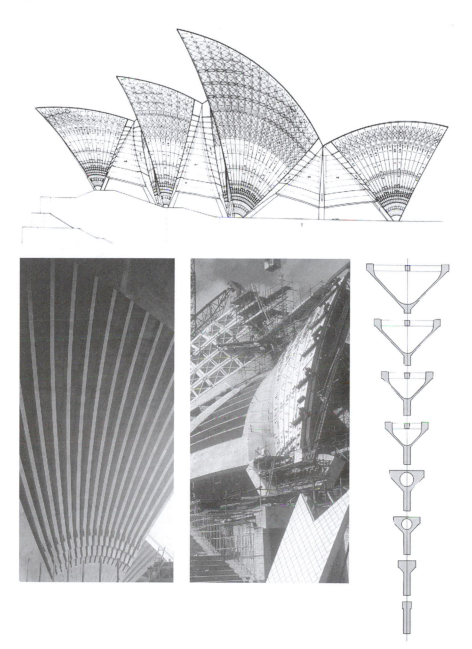

APPLICATION: PROPERTIES OF CROSS-SECTIONS: NATIONAL STADIUM, BEIJING (SEE PHOTOGRAPHS OPPOSITE)

The main stadium for the 2008 Beijing Olympics, or the 'Bird's Nest' as it became known, has an underlying primary structure of 24 'portalised trusses' which intersect to create a 3D space truss, up to 12 m deep between top and bottom chords. Typical steel box section members are 1.2 m × 1.2 m with wall thickness between 8 mm and 100 mm, with the majority below 70 mm.

Truss segments experience bending moments (from cladding loads between truss nodes) as well as axial forces, so section properties like second moment of area and section modulus introduced in this chapter, were crucial in assessing stresses arising from these internal actions, and the wall thicknesses necessary to receive those stresses. Wall thickness changes meant changes in dead load of the structure, hence the need to reanalyse the structure for revised internal actions. So the structural design was very much an iterative process, assisted greatly by pre- and post-processing routines attached to the analysis software. Prototype testing of joints to check joint computer modelling predictions was carried out at both Tsinghua University, Beijing, and Tongji University, Shanghai.

Project: Beijing National Stadium; Architect: Herzog and de Meuron; Structural Engineer: Arup with CADG (LDI); Photos: Arup

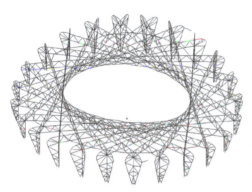

9 Bending Stresses

BENDING STRESS DISTRIBUTION

Consider a small section of a beam and identify two planes m and n a unit distance apart. Apply a uniform Bending Moment M to this section of the beam. The planes m and n remain plane (assumption 2) but are now inclined at an angle to one another.

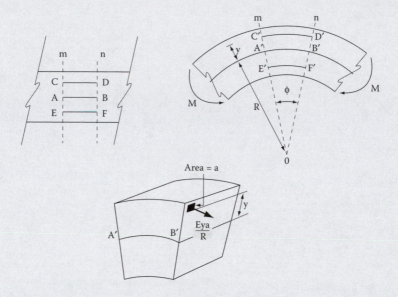

A fibre CD in the beam which originally was of unit length has increased to C'D'. Fibre EF has contracted to E'F'. There will be a fibre AB at some point which has not changed in length, i.e. A'B' also equals unit length. This fibre lies in what is termed the *neutral surface* of the beam: the surface that neither extends nor contracts during bending. Let the distance of the neutral surface from the centre of curvature (0) be R.

CONTENT OF CHAPTER 9

In Chapter 2 we saw how to calculate the stresses that exist in ties and struts. Under pure tension or pure compression, the stresses are uniform across the cross-section and are easily calculated. In beams resisting bending moments, the stresses are not uniform. They vary from tension to compression across the section, and usually along the beam as well.

In this chapter we will study the way bending stresses are distributed, introduce an important formula to calculate these stresses, and apply the idea of "section modulus" introduced in Chapter 8.

9.1 INTERNAL EFFECTS

In paragraph 7.4 we saw that the Bending Moment at a point in a beam is important because it measures the *total bending effect* produced at that point by the *external* forces. This bending effect produces a system of *internal* forces in the beam; in paragraph 1.13 we saw that these internal forces are tension near one side of the beam and compression near the other.

We may look at this matter another way. Suppose that at a plane m–m through a beam, the external forces have produced a Bending Moment M. Cut the beam through at m–m, and imagine what forces must be applied at the cut face to maintain equilibrium of the left-hand portion. (These forces would normally be applied by the right-hand portion, and are *internal* forces in the whole beam, as shown in Figure 9.1.)

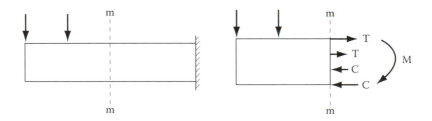

FIGURE 9.1

These, then, are the internal forces necessary to resist the applied Bending Moment M. Some of the forces will be tension and some compression, and their magnitudes would also be expected to vary across the section m–m. However, for each small tensional force we would expect to find a corresponding compressive force, and those two forces together would constitute a couple, producing a moment. The total *resistance* to bending of these forces at the section m–m would therefore be the sum of all

BENDING STRESS DISTRIBUTION   (CONTINUED)

Suppose that the distance of fibre CD from the neutral surface is y. Then,

change in length of fibre CD $= C'D' - CD$

$$= C'D' - AB$$

$\therefore$ strain in fibre CD $= \dfrac{C'D' - AB}{AB}$

$$= \dfrac{C'D'}{A'B'} - 1$$

$$= \dfrac{R + y}{R} - 1$$

$$= \dfrac{y}{R}$$

$\therefore$ stress in fibre CD $= f$

$$= \dfrac{E.y}{R}$$

$$\text{i.e.} = \dfrac{f}{y} = \dfrac{E}{R} \tag{1}$$

where E is the Young's Modulus of the material of the beam. Consider an infinitesimal area 'a' in plane m distant y from the neutral surface. The force on this area is then:

$$\text{stress} \times \text{area} = f.a$$

$$= \dfrac{Eya}{R}$$

However, for entire cross-section m, the total horizontal force must be zero, since there is no *external* horizontal force acting:

$$\text{i.e. } \Sigma f.a = 0$$

$$\therefore \dfrac{E}{R} \Sigma y.a = 0$$

$\therefore \Sigma y.a = 0$, and the *neutral surface must pass through the centroid of the cross-section* (paragraph 8.2).

these internal force couples. To satisfy the equilibrium of the left-hand portion of the beam, this *internal resistance moment* must equal the Bending Moment M caused by the *external* forces.

9.2 THE STRESS FORMULA

When one considers that people have been using beams in their dwellings and bridges for a period longer than all recorded history, it is interesting that the theory for the distribution of stresses was not developed until 1826. Navier, a French engineer, based his theory upon three principal assumptions, as follows:

1. Strain is proportional to stress for the material of the beam.
2. Cross-sections that are plane and normal to the beam axis before bending remain plane and normal after bending.
3. The beam is assumed to bend in a plane containing the vertical centroidal axis and the beam axis.

By considering the beam to be composed of a bundle of 'fibres' parallel to the axis of the beam (as in a solid timber beam), he was able to derive a relationship between applied bending moment, the properties of the beam, and the internal stress caused by the Bending Moment:

$$\frac{f}{y} = \frac{M}{I} = \frac{E}{R}$$

where
  f = bending stress (tension or compression in a particular fibre)
  y = distance of fibre from centroid of cross-sectional area
  M = applied B.M. at the cross-section being considered
  I = Second Moment of Area of the cross-section
  E = Young's Modulus of the material
  R = Radius of Curvature of the beam

The derivation of this relationship is shown opposite.

9.3 DEDUCTIONS

There are several very important conclusions that may be drawn from this derivation:

(i)   In every beam there will be a *neutral surface* (i.e. a surface within the beam that undergoes neither tensile nor compressive straining) that passes through the centroids of all cross-sections of the beam. For any one cross-section the intersection of the neutral surface and the plane of the cross-section is called the *neutral axis*. The neutral axis is thus a centroidal axis (Figure 9.2).

BENDING STRESS DISTRIBUTION   (CONTINUED)

The *moment* produced by the force in fibre CD about the centroid is then $\dfrac{Eya}{R}.y,$ and the sum of these internal moments is then:

$$\Sigma \frac{Eay^2}{R} = \frac{E}{R}\Sigma ay^2$$

But Σay^2 is the Second Moment of Area, I.

∴ Internal moment of resistance

$$= \frac{EI}{R}$$

For equilibrium, this internal moment equals the Bending Moment M.

$$\text{i.e. M} = \frac{EI}{R} \tag{2}$$

Combining Equations (1) and (2), we have

$$\frac{M}{I} = \frac{E}{R} = \frac{f}{y} \tag{3}$$

(ii) The magnitude of the bending stress at any fibre of the beam is directly proportional to the distance of the fibre from the neutral axis (Equation 1). The maximum bending stress occurs in the so-called *extreme fibre* (the fibre most remote from the neutral axis). Therefore the bending stresses vary linearly from a maximum tension on one side of the

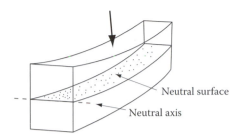

Neutral surface

Neutral axis

FIGURE 9.2

beam, through zero at the neutral axis, to a maximum compression on the other side of the beam. We can represent this as a graph (a stress diagram or stress block) showing variation of stress from top to bottom of the beam (Figure 9.3).

(iii) By rearrangement of Equation 3, we have:

$$f = \frac{My}{I}$$

i.e. the stress in any fibre is proportional to the bending moment and inversely proportional to the second moment of area of the beam cross-section.

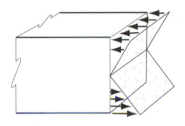

FIGURE 9.3

9.4   STRESS VARIATION

These three conclusions are of great importance in our understanding of behaviour of beams. The equation:

$$f = \frac{My}{I}$$

is a fundamental relationship in the study of flexure, and should be committed to memory. From it we see that, for beams of constant cross-section, the most highly stressed points are the extreme fibres at the location of maximum bending moment, because at these points both M and y are maxima. The stresses become lower as we consider fibres closer to the neutral axis or where the bending moments are smaller. Thus the magnitude of bending stress varies continuously throughout the beam, as shown in Figure 9.4. Each beam will have a unique distribution of stress depending

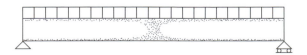

FIGURE 9.4

EXAMPLE 9.1

A timber beam is 100 mm wide and 200 mm deep. What is the stress at each of the following points at a section where the Bending Moment is 8 kNm?

(i)   at the top surface
(ii) 50 mm below the top surface
(iii) 100 mm below the top surface
(iv) 150 mm below the top surface

$$I = \frac{bd^3}{12} = \frac{100 \times 200^3}{12} = \frac{800}{12} \times 10^6$$

$$= 66.7 \times 10^6 \text{ mm}^4$$

$$\text{B.M.} = 8 \text{ kNm} = 8 \times 10^6 \text{ Nmm}$$

The Neutral Axis is at the centroid of the section (i.e. 100 mm from the top or bottom).

(i) $f = \dfrac{M.y}{I} = \dfrac{8 \times 10^6 \times 100}{66.7 \times 10^6}$ $= 12$ N/mm^2

$= 12$ MPa (compression)

(ii) $f = \dfrac{M.y}{I} = \dfrac{8 \times 10^6 \times 50}{66.7 \times 10^6}$ $= 6$ N/mm^2

$= 6$ MPa (compression)

(iii) $f = \dfrac{M.y}{I} = \dfrac{8 \times 10^6 \times 0}{66.7 \times 10^6}$ $= 0$ MPa

(iv) $f = \dfrac{M.y}{I} = \dfrac{8 \times 10^6 \times (-50)}{66.7 \times 10^6} = -6$ N/mm^2

$= 6$ MPa (tension)

upon its cross-section and upon the way it is supported and loaded. The equation given in this paragraph defines that distribution for all beams.

9.5 MAXIMUM STRESS

Although it is of value to us to be able to determine the bending stress at *any* point in a beam (see Example 9.1), it is *especially* important for us to be able to calculate quickly the *maximum* fibre stress in the loaded beam. We saw in Chapter 2 that the Permissible Stress Design method requires that the maximum stress in a structure does not exceed the permissible stress. The maximum stress will be:

$$f \, \text{max} = \frac{M \, \text{max}.c}{I}$$

where c is the maximum value of y, i.e. c is the maximum distance from the neutral surface to the extreme fibre at the location of maximum Bending Moment.

In paragraph 8.4 we saw that I/c is called the Section Modulus of the cross-section, and is given the symbol Z. Hence:

$$f \, \text{max} = \frac{M \, \text{max}}{Z}$$

This relationship is also very important and should be committed to memory. It enables us to calculate very easily the *maximum bending stress* (linear elastic) in any beam. For most of the beam shapes that we use in buildings, the values of Z have been computed and are available in tabular form or are stored in computer programs. To find the maximum stress we merely have to evaluate the maximum Bending Moment and divide by this pre-calculated Z.

A brief mention should be made of non-symmetrical cross-sections. In paragraph 8.4, we noted that if the cross-section is not symmetrical, there will be two values of Z. Obviously the maximum bending stress will be calculated on the minimum value of Z.

$$f \, \text{max} = \frac{M \, \text{max}}{Z \, \text{min}}$$

Figure 9.5 shows stress diagrams for some typical cross-sections.

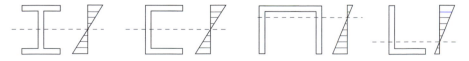

FIGURE 9.5

WORKSHEET 9.1

9.1 For the beam shown in Figure 1, calculate the bending stresses in the fibres located at the top of the beam, 25 mm from the top, 50 mm from the top and 150 mm from the top, at each of the sections AA and BB. The beam is of rectangular cross-section, 100 mm wide × 200 mm deep.

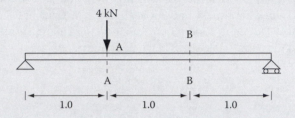

FIGURE 1

9.2 A beam having the cross-section shown in Figure 2a is loaded as shown in Figure 2b. The values of the Section Modulus for the beam are 1.74×10^6 and 0.74×10^6 mm^3 (see Question 8.2). What are the extreme fibre stresses at sections B and C?

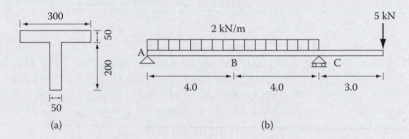

FIGURE 2

9.3 Draw a graph showing the distribution of bending stress across each of the cross-sections in Questions 9.1 and 9.2 above (refer to Figure 9.3).

9.4 In Figure 3 are shown five beam cross-sectional shapes, *all having the same area*. For each of these, sketch the distribution of bending stress so as to indicate the relative magnitude of stresses.

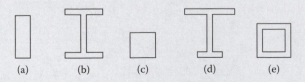

FIGURE 3

9.6 'ULTIMATE' CONDITIONS

The linear elastic bending stresses described above apply for most materials up to a certain level of bending. At higher Bending Moments, the linear stress distribution of Figure 9.3 needs modification to reflect the real stress/strain behaviour of the particular material.

For example, in a steel beam, the extreme fibre stress will stop increasing when it reaches the yield stress for the steel, in compression and in tension. It is possible, however, to keep increasing the applied Bending Moment until nearly all the fibres above the neutral axis have yielded in compression, and those below in tension. At this point, the stress distribution is mainly rectangular, not triangular, and the beam is said to have reached its *ultimate moment capacity*. In reinforced concrete, as a beam approaches its ultimate capacity, the stress distribution on the compression side is more curved than linear.

Knowing the ultimate strength capacity of a cross-section in bending allows use of the Ultimate Strength Design method described in Chapter 2. Because this method is based on the actual beam strength and reflects real material behaviour at high stresses more closely, it is generally favoured over the Permissible Stress Design method, where no portion of the cross-section is allowed to yield. These methods are considered further in paragraph 12.4.

APPLICATION: BEAMS IN BENDING, PHOENIX CENTRAL LIBRARY
(SEE PHOTOGRAPHS OVER PAGE)

Precast, prestressed concrete structures, as used for the Phoenix Central Library, Arizona, often feature beam elements spanning simply-supported between bearings. The BMD and SFD methods of Chapter 7, the cross-section property analyses of Chapter 8, and the stress analyses of Chapter 9, are directly applicable in such structures. At Phoenix, the precast 'double T' floor planks span onto the precast, rectangular cross-section primary beams or 'girders', which bear in turn on the column corbels. The voids between the webs of the planks then allow passage of services in the completed building. The exposed structural concrete surfaces contribute to the building's thermal inertia and hence its strategy for energy efficiency, which benefits from night-time air-flushing in the ideal desert climate, and also the long occupancy hours of the library.

Project: Phoenix Central Library, Phoenix, Arizona, USA; Architect: Bruder DWL Architects; Structural Engineer: Arup; Photos: (top, middle, bottom left): Bill Timmerman; (bottom right): Arup

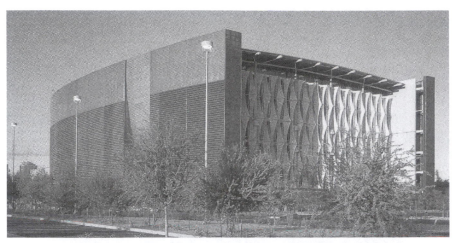

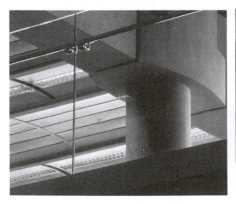

10 Shear Stresses

SHEAR STRESS DISTRIBUTION IN A RECTANGULAR BEAM

Consider a rectangular beam of width b and depth d, and identify two transverse sections m and n a distance x apart. Let the bending moment at m be M_1 and at n be M_2.

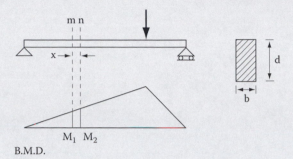

B.M.D.

Isolate the small portion of the beam between planes m and n and apply to it the bending stresses produced by the moments M_1 and M_2.

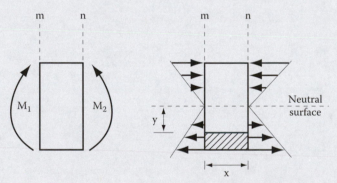

Now consider a horizontal plane distant y from the neutral surface and consider the horizontal equilibrium of the portion of the beam below this surface.

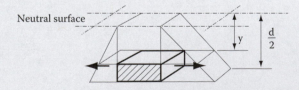

The average bending stress on the right-hand face of this block is:

$$f_2 = \frac{M_2 \bar{y}}{I},$$ where $\bar{y}$ is the distance of the centroid of his face from the neutral surface.

CONTENT OF CHAPTER 10

In paragraph 9.1 we saw that Bending Moment produces bending stress in a beam. Likewise Shear Force produces shear stress in the beam — stresses that resist the tendency of the left-hand part of the beam to slide vertically with respect to the right-hand part.

We know that the average shear stress is found by dividing the Shear Force by the cross-sectional area of the beam:

$$fs(aver) = \frac{V}{A}$$

In this chapter, we will see how the shear stress varies at different locations in the beam; where the shear stress will be a maximum, and by how much the maximum stress exceeds the average stress.

10.1 HORIZONTAL AND VERTICAL SHEAR STRESS

In Chapter 7 we saw how Shear Force varies along a beam, while in paragraph 2.3 we studied the way shear stresses are produced in small structural members such as bolts. We might infer from these descriptions that shear stresses in beams are *vertical* stresses that prevent a *vertical* slippage from occurring.

In fact, beams rarely if ever fail by one part sliding vertically with respect to the other. This is because, as well as the *vertical or transverse* shear stresses we have already encountered, there are also *horizontal or longitudinal* shear stresses in a beam under load. There are two ways we can demonstrate this:

(a) If we take a number of planks, support them at the ends and apply a load at midspan, the planks deflect as shown (Figure 10.1).

FIGURE 10.1

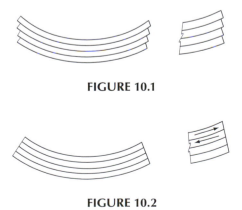

FIGURE 10.2

SHEAR STRESS DISTRIBUTION IN A RECTANGULAR BEAM (CONTINUED)

Thus if the area of this face is A, the total *force* on this area is:

$$F_2 = \frac{M_2\bar{y}}{I}.A$$

Similarly, the force on the left-hand face is:

$$F_1 = \frac{M_1\bar{y}}{I}.A$$

Thus the horizontal forces acting on the block are as shown below, and these cannot be in equilibrium since $M_2 > M_1$ i.e. there must be another horizontal force acting on the block to restore equilibrium, and this unknown force must be a shear force, F_S acting on the upper face.

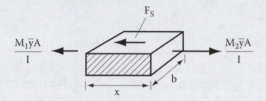

Thus $$\frac{M_2\bar{y}A}{I} - \frac{M_1\bar{y}A}{I} - F_S = 0$$

i.e. $$F_S = \frac{(M_2 - M_1)\bar{y}A}{I}$$

The shear stress acting on the upper face is F_S divided by the area. Let b be the width of the beam. Then the shear stress f_S is given by:

$$f_S = \frac{(M_2 - M_1)A\bar{y}}{xbI}$$

But, as $x \to 0$, $\dfrac{M_2 - M_1}{x}$ becomes the *slope* of the Bending Moment Diagram, and this equals the value of Shear Force at this section as read from the Shear Force Diagram (refer paragraph 7.6).

Let the value of the Shear Force be V. Then:

$$f_S = \frac{VA\bar{y}}{Ib}$$

Each plank acts as a separate beam and independently carries its share of the load. If we removed the load, and glued the planks together, we would, in effect, have a solid beam of the same overall dimensions as the bundle of planks, but when *this* beam is loaded it will deflect in a different way. The glue has prevented the bottom surface of each plank from sliding lengthways over the plank below. Thus, in a solid beam there must be shear forces which act *parallel* to the neutral surface to prevent lengthwise slippage of the fibres in the beam (Figure 10.2).

(b) If we consider a beam carrying loads and imagine what forces need to be applied to maintain equilibrium as we cut the beam through at two adjacent sections (m and n, Figure 10.3), we see that the left-hand part must apply an *upward* force to the right-hand part, and the right-hand part must apply a *downward* force to the left-hand part. If we now consider an infinitesimally small cube of material cut from the beam between the sections m and n, we see that the two shear forces acting on the vertical faces produce a clockwise couple. Since this cube is in equilibrium when it is in the beam, the beam must also be applying an anticlockwise couple to the cube. So there must be a pair of *equal horizontal* shear forces acting on the horizontal faces of the cube. Thus, in a beam there must be shear forces that act both parallel and perpendicular to the neutral surface, and at any particular point these stresses are all equal.

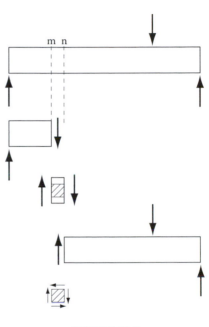

FIGURE 10.3

Consequently, we would expect to find *two* sets of shear stresses within any beam, and the resistance of the beam to these stresses will prevent shear slippage from occurring both vertically and horizontally.

10.2 SHEAR STRESS FORMULA

On the opposite pages an expression for shear stress has been derived for a rectangular beam. It is:

$$f_s = \frac{VA\bar{y}}{Ib}$$

where:

f_s is the longitudinal shear stress acting on a particular plane

V is the external Shear Force acting

EXAMPLE 10.1

A rectangular beam of width 200 mm and depth 300 mm supports loads producing a Shear Force of 300 kN at a particular section. What shear stresses exist at the following points?

(i)   100 mm from the top
(ii) at mid-depth
(iii) at the bottom

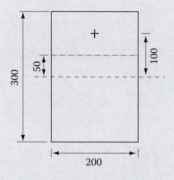

$$f_s = \frac{VA\bar{y}}{Ib}$$

$V = 300 \text{ kN} = 300 \times 10^3 \text{ N}$

$b = 200 \text{ mm}$

$$I = \frac{bd^3}{12} = \frac{200 \times 300^3}{12}$$
$$= 450 \times 10^6 \text{ mm}^4$$

(i) $A = 200 \times 100 = 20 \times 10^3 \text{ mm}^2$

$\bar{y} = 100 \text{ mm}$

$$\therefore f_s = \frac{(300 \times 10^3) \times (20 \times 10^3) \times 100}{450 \times 10^6 \times 200}$$
$$= 7 \text{ N/mm}^2 = 7 \text{ MPa}$$

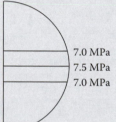

7.0 MPa
7.5 MPa
7.0 MPa

(ii) $A = 200 \times 150 = 30 \times 10^3 \text{ mm}^2$

$\bar{y} = 75 \text{ mm}$

$$\therefore f_s = \frac{(300 \times 10^3) \times (30 \times 10^3) \times 75}{450 \times 10^6 \times 200}$$
$$= 7.5 \text{ MPa}$$

Shear stress distribution
over depth of beam

(iii) $A = 0$

$\therefore f_s = 0$

A is the area of that part of the beam cross-section between the particular plane and
 the extreme fibre

$\bar{y}$ is the distance from the neutral surface to the centroid of A

I is the second moment of area of the beam cross-section

b is the width of the beam at the particular plane.

Although we have derived this expression for a rectangular beam, it is true for all
cross-sections. The terms in this expression are more easily appreciated if they are
related to a particular beam cross-section, as in Figure 10.4.

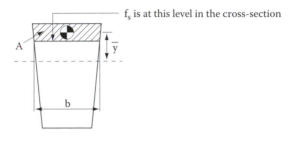

FIGURE 10.4

10.3   MAXIMUM SHEAR STRESS

The way this expression may be used to compute shear stresses in a rectangular
beam is shown in Example 10.1. From this example, we see that the maximum shear
stress occurs at the neutral axis. At this point, $A = bd/2$, $\bar{y} = d/4$, and we have:

$$f_s \, max = \frac{VA\bar{y}}{Ib} = \frac{V.bd.d}{8b} \times \frac{12}{bd^3} = \frac{3}{2}\frac{V}{bd}$$

Now bd is the cross-sectional area of the beam, and V/bd is the *average* shear stress,
as discussed in Chapter 1. Thus the maximum shear stress *in a rectangular beam* is
1.5 times the average shear stress. This result should be carefully noted.

Example 10.2 shows the application of the expression to an I-beam. This example
should also be noted, since it shows well the effects of the variables A, $\bar{y}$ and b.

For an *I-beam,* we see that the shear stresses in the flanges are usually very small,
whereas those in the web are very large. In fact, most of the shear strength of an
I-beam comes from the web. Moreover, although the shear stress is still a maximum
at the neutral axis, the variation across the depth of the web is quite small, i.e. the
shear stress in the web is approximately constant. In fact, steel beams are usually
designed as though the web shear stress were constant, and the shear stress is taken
to be the shear force divided by the overall depth of the beam times the thickness of
the web.

EXAMPLE 10.2

For the I-section shown, calculate the shear stress at the junction of the web and the flange, and at the neutral axis, when the Shear Force is 200 kN.

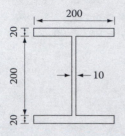

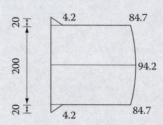

Distribution of shear stress across the I-beam.

$$I = \frac{200 \times 240^3}{12} - \frac{190 \times 200^3}{12} = 104 \times 10^6 \text{ mm}^4$$

At top of web:

$$A = 200 \times 20 = 4 \times 10^3; b = 10; \overline{y} = 110$$

$$f_s = \frac{200 \times 10^3 \times 4 \times 10^3 \times 110}{104 \times 10^6 \times 10} = 84.7 \text{ MPa}$$

At mid-depth:

$$A = 4 \times 10^3 + 10 \times 100 = 5 \times 10^3; b = 10$$

$$\overline{y} = \frac{\Sigma ay}{A} = \frac{4 \times 110 \times 10^3 + 1 \times 50 \times 10^3}{5 \times 10^3}$$

$$= 98 \text{ mm}$$

$$f_s = \frac{200 \times 10^3 \times 5 \times 10^3 \times 98}{104 \times 10^6 \times 10} = 94.2 \text{ MPa}$$

At bottom of top flange:

$$A = 4 \times 10^3; b = 200; \overline{y} = 110$$

$$f_s = \frac{200 \times 10^3 \times 4 \times 10^3 \times 110}{104 \times 10^6 \times 200} = 4.2 \text{ MPa}$$

10.4 DEDUCTIONS

We are now able to draw a number of general conclusions about the distribution of shear stresses in a beam:

(i) Shear stress is directly proportional to Shear Force.
(ii) Shear stress is always zero at the top and bottom fibres of a beam (because A is zero).
(iii) Shear stress is always a maximum at the neutral axis (because the product $A\bar{y}$ is a maximum).
(iv) Shear stress varies parabolically over parts of a beam for which the thickness b is constant (because the product $A\bar{y}$ varies parabolically).

Permissible Stress Design, outlined in Chapter 2, allows us to specify a maximum permissible value for the maximum shear stress in (iii), depending on the material in the beam. This allows us to design the beam cross-section to be strong enough in shear. If we consider the distribution of shear stress in a simply supported beam carrying a uniformly distributed load, we find that the shear stress is a maximum at mid-depth near the supports, and zero along top and bottom fibres and at mid-span (Figure 10.5). The shear stresses are at a maximum at points where the bending stresses are zero, and vice versa.

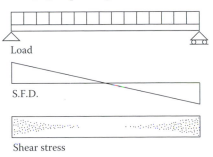

Load

S.F.D.

Shear stress

FIGURE 10.5

It is interesting to note that at the mid-depth midspan point, the fibres are quite stress-free. This suggests that if, for any reason, it is necessary to cut a fairly large hole through a beam (for example, for a duct or pipe to pass through), then the mid-depth, midspan location is an obvious choice.

We have seen that in an I-beam, the flanges do most of the work in bending and have high bending stresses, whereas the web provides most of the shear strength. An I-beam is an economical section because the thin web saves material yet is still able to provide the requisite shear strength. We could, of course, also save material by having a thicker web, pierced by large holes at fairly regular intervals (Figure 10.6), yet still retain the same average web area and hence (approximately) the same shear strength.

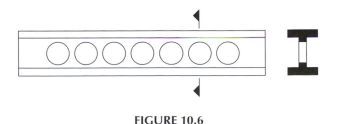

FIGURE 10.6

WORKSHEET 10.1

10.1 A rectangular timber beam 50 mm wide by 100 mm deep is formed by glueing together two pieces of timber each 50 mm square. The beam is supported between two walls 2 m apart and a central point load of 2 kN is applied. What is the maximum bending stress in the timber? What is the maximum shear stress at the glue-line?

10.2 If we had simply *rested* one piece of timber on the other, without glueing them together as in Question 10.1, what would have been the maximum bending stress in the timber for the loadings given in Question 10.1? What can you deduce from this?

10.3 For the beam shown in Question 9.2:

    (a) What is the shear stress at the junction of the web and the flange near point A?
    (b) What is the maximum shear stress near point A?
    (c) What is the maximum shear stress at *any* point in the beam?

10.4 For each of the five beam cross-sectional shapes shown in Question 9.4, sketch the distribution of shear stress so as to indicate the relative magnitude of stress under the action of a given shear force.

Such beams often have hexagonal holes from the manufacturing process, and are called castellated beams. If we imagine the holes to be quite large and triangular, we have a truss, as in Figure 10.7. In fact, a truss behaves very similarly to a beam, with the upper and lower members (the *chords*) providing most of the bending resistance (like flanges) and the web members providing the shear resistance. The stresses in the chords (like flanges) are large at midspan and small at the ends. The stresses in the web members (like shear stresses in a web) are small at midspan and large at the ends. The advantage of a truss is that we do not need to use the same section throughout the whole length, but can use heavier chords and web members where the internal forces are larger.

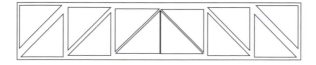

FIGURE 10.7

An 'ideal' beam would be one having an I section at midspan (for bending), but a round, square or rectangular section at the ends to keep the shear stresses low. Beams of varying cross-section can be extremely efficient, although usually expensive to manufacture.

In the manufacture of large, glued-laminated timber beams ('glulam' beams), particular care is taken to select the best pieces of timber for the outermost laminations, because these are the ones that will be most highly stressed in bending. Poorer-quality timber can be used near the centre, where (usually smaller) shear stresses need to be resisted. In a similar way, 'stressed skin' cladding panels are fabricated so as to have a pair of strong, stiff skins on the outside (e.g. plywood or fibreglass), with a relatively weaker material (e.g. foamed polystyrene) as the core. The fibreglass provides the bending strength and the polystyrene provides the shear strength.

10.5 'ULTIMATE' CONDITIONS

The theory outlined above assumes linear elastic stress/strain behaviour of the beam material. This applies for most materials up to a certain level of shear. At higher shear, local yielding (or local cracking in reinforced concrete) causes some parts of the cross-section to reach their stress capacity. Any further applied shear force is then carried by other parts of the section until it eventually fails in shear (or shear plus bending) at its *ultimate strength*.

Knowing the ultimate strength capacity in shear allows use of the Ultimate Strength Design method described in Chapter 2 and considered further in Chapter 12.

APPLICATION: DESIGN FOR SHEAR, CUMMINS ENGINE FACTORY
(SEE PHOTOGRAPHS OPPOSITE)

It is the web of a steel 'I' section beam that carries most of an applied shear force. 'Castellated' beams can be an economical choice when shear stresses in a given application are low relative to bending stresses, so that a significant portion of the web of the beam can be deleted. This is achieved by a zig-zag cut along the web of a conventional 'I' beam, separation and translation of the two half-beams, then welding back together at the matching web plate edges.

At the Cummins diesel engine factory in Lanarkshire, Scotland (above), long spans and light roof loading led to a high bending, high deflection, low shear situation for the roof beams. Castellation provided a deeper beam for the same steel tonnage, hence lower bending stresses due to the increased section modulus, and lower deflections due to the increased second moment of area (see Chapter 11).

Project: Cummins Engine Company, Shotts, Lanarkshire, Scotland; Architect: Ahrends, Burton, Koralek; Structural Engineer: Arup; Photos: Arup

APPLICATION: DESIGN FOR SHEAR, SERPENTINE PAVILION 2005
(SEE PHOTOGRAPHS OPPOSITE)

It is unusual for the size of structural members to be governed by shear strength considerations. The 'lamella' structure chosen for the 2005 Serpentine Pavilion in Hyde Park, London, is one such example. The grid roof comprises laminated veneer lumber (LVL) lamellas or blades, each with end tenons or tongues that fit into mortices or slots in the adjacent lamellas.

Each lamella receives a large point load at its mid-length from the incoming tenons, and is supported at its ends by its own tenons bearing in neighbouring mortices. The depth of the lamellas is determined by the need to deliver a significant shear force through the tenons, and receive it on the lamella mid-span cross-section which has reduced strength because of the presence of the mortice. Each lamella was defined by 36 set-out parameters. A spreadsheet of these parameters provided direct input to the CNC 6-axis robot that cut the LVL pieces, without the need for fabrication drawings.

Project: 2005 Serpentine Pavilion, Hyde Park, London, UK; Architect: Alvaro Siza and Eduardo Souto de Moura; Structural Engineer: Arup; Photos: Arup

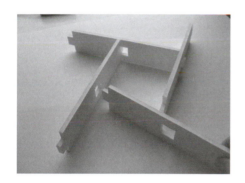

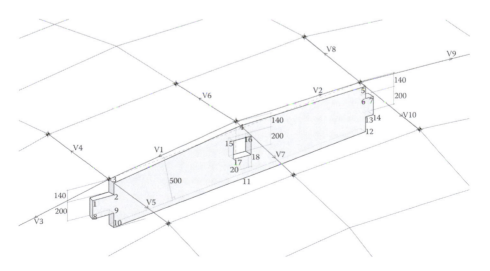

11 Deflections

VALUES FOR SECOND MOMENT OF AREA

The Second Moment of Area of the cross-sectional shape of a beam has a direct influence upon the amount that the beam deflects. The following table gives formulae for some common geometrical shapes.

TABLE 11.1

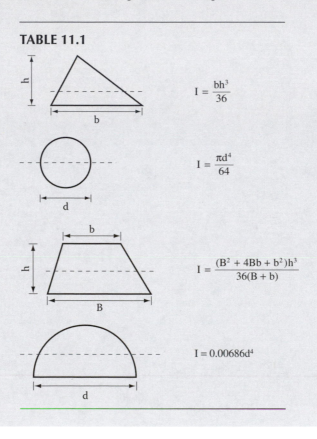

$$I = \frac{bh^3}{36}$$

$$I = \frac{\pi d^4}{64}$$

$$I = \frac{(B^2 + 4Bb + b^2)h^3}{36(B + b)}$$

$$I = 0.00686d^4$$

CONTENT OF CHAPTER 11

This chapter deals with the deflection of beams. After deducing the general form of the deflection equation, we consider its individual components, to understand the influence that each has on the overall deflection. The relative effects of different load configurations and support conditions are also considered. Mention is made of the parameters affecting dynamic, rather than just static, deflections.

11.1 GENERAL RULES

It is quite a simple matter to determine experimentally the factors that influence the amount of deflection that a structure undergoes when a load is applied to it. If a simple beam made from a linear elastic material is subjected to a central point load, it will be found that:

(a) the maximum deflection is proportional to the load;

(b) the maximum deflection is proportional to the cube of the span;

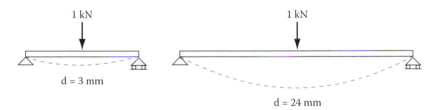

DEFLECTION OF A CANTILEVER

The maximum deflection of a beam depends, as we see in paragraph 11.1, on load, span, cross-sectional shape and material of construction.

Consider a cantilever beam supporting a point load at the free end. The beam will deflect to a smooth curve, with each small increment of length bending through a small angle.

Consider one such increment of length, dx, bending through an angle θ. The angle between the perpendiculars to the beam will also be θ, and the radius of curvature is R. Then:

$$\theta = \frac{dx}{R}$$

In Chapter 9, we saw that:

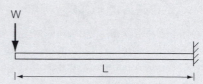

$$\frac{M}{I} = \frac{E}{R} = \frac{f}{y}$$

i.e. $$\frac{1}{R} = \frac{M}{EI}$$

$$\therefore \theta = \frac{Mdx}{EI}$$

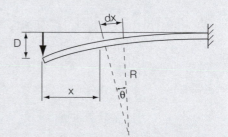

If the increment dx is located a distance x from the free end:

$$M = W.x \quad \text{and} \quad \theta = \frac{W.x.dx}{EI}$$

Let the deflection at the free end caused by the bending in the short length dx be d.

$$\therefore d = x.\theta = \frac{W.x^2.dx}{EI}$$

Thus the total deflection D at the free end is the sum of the values of d for all increments of length between x = 0 and x = L,

$$\text{i.e. Deflection D} = \int_{O}^{L} \frac{W.x^2.dx}{EI}$$

$$= \frac{1 WL^3}{3\ EI}$$

Thus, for the case of a cantilever loaded at its free end, the deflection is proportional to the load and to the cube of the span, and inversely proportional to the Second Moment of Area (i.e. shape) and the Young's Modulus (i.e. material stiffness).

(c) the maximum deflection depends upon the cross-sectional shape of the beam;

(d) the maximum deflection depends upon the material.

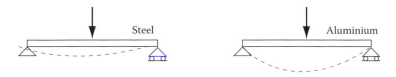

11.2 DEFLECTION FORMULA

In fact, the maximum deflection of a beam is given by the following general expression:

$$\text{Maximum Deflection} = C\frac{WL^3}{EI}$$

where

    C = a constant
    W = total load
    L = span
    E = Young's Modulus of Elasticity of the material of the beam
    I = Second Moment of Area of the cross-section of the beam.

The constant C depends upon the way in which the beam is supported and also upon the pattern of loads that are applied to it. Table 11.2 shows the value taken by this constant for a number of commonly encountered conditions of support and load. For a cantilever of length L carrying a point load of W at its free end, we see from this table that the maximum deflection will occur at the free end, and will have a value of:

$$\frac{1}{3}\frac{WL^3}{EI}$$

For a simply supported beam of span L carrying a total uniformly distributed load of W, the maximum deflection will be at midspan, and will have a value of:

$$\frac{5}{384}\frac{WL^3}{EI}$$

It is important to use the correct *units* with this formula. Because E will usually be quoted in MPa, all other terms will need to be in *newtons and millimetres*. Thus

EXAMPLE 11.1

Calculate the maximum deflection for the steel beam loaded as shown, given:

$$I = 20 \times 10^6 \text{ mm}^4$$

$$E = 210 \times 10^3 \text{ MPa}$$

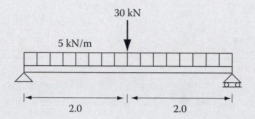

Deflection due to point load:

$$= \frac{1}{48} \frac{WL^3}{EI}$$

$$= \frac{1 \times 30 \times 10^3 \times 4^3 \times 10^9}{48 \times 210 \times 10^3 \times 20 \times 10^6} = 9.5 \text{ mm}$$

Deflection due to U.D. load:

$$= \frac{5}{384} \frac{WL^3}{EI}$$

$$= \frac{5 \times 20 \times 10^3 \times 4^3 \times 10^9}{384 \times 210 \times 10^3 \times 20 \times 10^6} = 4.0 \text{ mm}$$

Total deflection = 13.5 mm

Note: This example illustrates an application of the Principle of Superposition: the deflection caused at a point by a complex load pattern is the sum of the deflections produced at that point by simpler load patterns considered separately, for a linear elastic structure.

a load in kN will need to be multiplied by 10^3. A span in metres would also be multiplied by 10^3 but, because we need the cube of span, we must multiply by 10^9. Example 11.1 illustrates the method.

For the sake of completeness, Table 11.2 also includes the value of the maximum Bending Moment.

It should be noted carefully that W in this formula ('capital' or 'upper case' W) is the *total* load shown in the appropriate diagram of Table 11.2. It will usually be expressed in kN. In other text books, you may encounter a 'little' or 'lower case' w, used in a formula involving L^4 e.g. for a simply supported beam carrying a uniformly distributed load, the deflection may be given as:

$$\frac{5}{384} \frac{wL^4}{EI}$$

In this usage, w is the *intensity* of a *distributed load,* and will be expressed in kN/m. Obviously, the two forms of expression are equivalent because the total load W (kN) is equal to the intensity of load w (kN/m) multiplied by the length L (m). Either form may be used.

11.3   SPANS AND MATERIALS

Of the various parameters which we have shown to affect deflection, the span length L is of the greatest importance, because the deflection is proportional to the span cubed. A 10% increase in span produces a 33% increase in deflection. A beam in a building spanning 6 m may have a midspan deflection of, say, 20 mm. If the span is increased by a mere 0.6 m, the deflection becomes 27 mm, and if the span is increased to 9.0 m, the deflection would be more than 67 mm!

So, if we wish to use long-span beams, we must increase the value of I, i.e. we must use deeper, heavier beams to reduce the deflections to acceptable values. If the span is doubled, then the Second Moment of Area must be increased by *eight times* to obtain the same deflection. We see that it becomes very costly in terms of materials to use long-span beams, and in fact there is an economic limit to the span length for different forms of construction. Thus we find that solid timber beams are less likely to be used for spans exceeding 6 m, reinforced concrete about 10 m, steel about 15 m, and prestressed concrete about 25 m. If we require spans of greater length than these, we usually abandon the use of solid beams altogether, and use other structural systems — trusses, arches, domes, folded plates, etc.

WORKSHEET 11.1

11.1 A beam having the cross-section shown in Question 9.2 is simply supported over a span of 3.5 m. If the material of the beam has a Young's Modulus of 10×10^3 MPa, what would be the maximum midspan deflection under each of the following conditions?

    (a) A uniformly distributed load of 2 kN/m.
    (b) A central point load of 7 kN.

11.2 The cross-sectional area of the beam in the previous question is 25 000 mm². If, instead of using this quantity of material to form a Tee-beam, we had used it to form a beam of equal area, but having a cross-section 158 mm square, what would the maximum deflection have been under the 7 kN point load? If the cross-section had been circular, what would have been the maximum deflection?

11.3 A hardwood beam of rectangular cross-section 100×250 is simply supported over a span of 4.8 m, and is loaded as sketched below. What would be its maximum deflection? ($E = 10.5 \times 10^3$ MPa.)

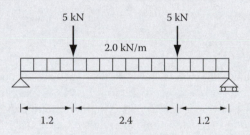

TABLE 11.2

Deflections and Bending Moments for Some Common Loading Conditions

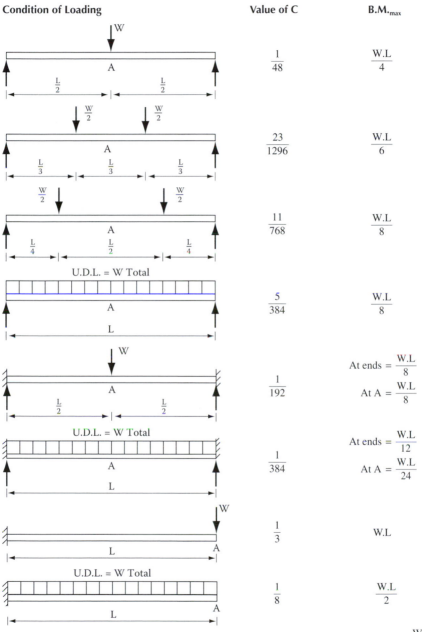

Condition of Loading	Value of C	B.M.$_{max}$
(central point load W, L/2 + L/2)	$\dfrac{1}{48}$	$\dfrac{W.L}{4}$
(two loads W/2 at L/3 intervals)	$\dfrac{23}{1296}$	$\dfrac{W.L}{6}$
(two loads W/2 at L/4, L/2, L/4)	$\dfrac{11}{768}$	$\dfrac{W.L}{8}$
U.D.L. = W Total (simply supported, span L)	$\dfrac{5}{384}$	$\dfrac{W.L}{8}$
(fixed ends, central load W, L/2 + L/2)	$\dfrac{1}{192}$	At ends $= \dfrac{W.L}{8}$ At A $= \dfrac{W.L}{8}$
U.D.L. = W Total (fixed ends, span L)	$\dfrac{1}{384}$	At ends $= \dfrac{W.L}{12}$ At A $= \dfrac{W.L}{24}$
(cantilever, end load W, length L)	$\dfrac{1}{3}$	W.L
U.D.L. = W Total (cantilever, length L)	$\dfrac{1}{8}$	$\dfrac{W.L}{2}$

Note: The maximum deflection occurs at the point marked A on each diagram, and has the value $C\dfrac{WL^3}{EI}$. The maximum value of B.M. also occurs at point A, except for the four fixed-ended examples, for which the maximum B.M. occurs at the fixed ends.

11.4   DEFLECTION SENSITIVITY

Deflection is a very common cause of trouble in building structures. Buildings rarely collapse, but every day we can see evidence of buildings showing distress resulting from deflection. For example, brick walls crack if the supporting beams deflect excessively and, although the beams may have abundant *strength* to carry their loads, the *stiffness* often is inadequate. The designer must always ensure that the structure and the claddings or finishes are compatible — that the stiffness of the structure is sufficient to prevent damage to the construction that the structure supports.

11.5   DYNAMICS

Another common cause of trouble in building structures in excessive vibration. This may be caused by people walking, running or dancing, by machinery oscillating, or by fluctuating wind pressure. All the same parameters involved in the deflection equation — span, load (or mass), Young's Modulus, Second Moment of Area, and conditions at the end of the beam — are important in calculating dynamic response, as well as the oscillating force input, and the damping available to help reduce the response.

APPLICATION: CONTROLLING DEFLECTIONS, COMMERZBANK
HEADQUARTERS (SEE PHOTOGRAPHS OPPOSITE)

For light steel-framed buildings, like the Commerzbank Headquarters in Frankfurt, deflections are checked at many locations and under many loadcases. Among the deflections studied here were the lateral movements of the top of the building under wind loads; the vertical deflection of the 8-storey tall facade bending frames that span 34 m between corner cores; and the floor beams that span 15.6 m between internal and external facade support structures.

The 560 mm deep steel beams lose some stiffness from the multiple services penetrations through their webs, and the notching of each end also to facilitate passage of services. They benefit however from 'composite action' with the 130 mm thick concrete floor slab, which acts as an additional top flange to the steel beams.

Project: Commerzbank Headquarters, Frankfurt, Germany; Architect: Foster and Partners; Structural Engineer: Arup; Photos: Ian Lambot; Diagram: Arup (Nigel Whale)

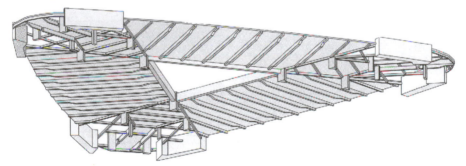

12 Design of Beams

EXAMPLE 12.1

Design a hardwood beam to support a uniformly distributed load of 14 kN over a simply supported span of 4 m. Consider bending strength only.

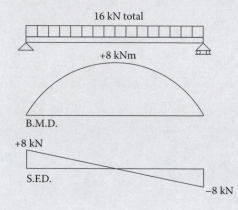

16 kN total

+8 kNm

B.M.D.

+8 kN

S.F.D.

−8 kN

Assume that the weight of the beam is 2 kN.

$\therefore$ Total load = 16 kN

$$\text{B.M.}_{\cdot max} = \frac{WL}{8} = \frac{16 \times 4}{8} = 8 \text{ kNm}$$

$$\text{S.F.}_{\cdot max} = 8 \text{ kN}$$

Permissible bending stress:

$$F_b = 11.0 \text{ MPa}$$

Permissible shear stress:

$$F_s = 1.0 \text{ MPa}$$

Moment capacity $M = F_b.Z$

i.e. we need $F_b.Z = 8 \times 10^6$ Nmm

$\therefore$ Required $Z = \dfrac{8 \times 10^6}{11} = 0.725 \times 10^6 \text{ mm}^3$

If beam is 75 mm wide:

$$Z = \frac{bd^2}{6} = \frac{75 \times d^2}{6} = 0.725 \times 10^6$$

$$\therefore d = \left(\frac{6 \times 0.725 \times 10^6}{75} \right)^{\frac{1}{2}}$$

$$= (0.058 \times 10^6)^{\frac{1}{2}} = 240 \text{ mm}$$

Thus, a beam of nominal dimensions 75×250 would be satisfactory for bending strength. Moreover, it will have adequate resistance to lateral instability because its depth is only a little more than three times its width.

CONTENT OF CHAPTER 12

This chapter completes our study of elementary structure, by bringing together most of the preceding concepts into a real application: the design of a simple timber beam. This involves combining our understanding of loads, reactions, bending moment and shear force, cross-section properties, bending stress and shear stress, and deflections, to test that conditions for the beam will be acceptable.

This provides us with a simple but powerful procedure for sizing key bending elements. Mention is also made of 'ultimate strength' methods, which rely on strength-at-failure rather than requiring limits on local stresses under working loads.

12.1   ANALYSIS AND DESIGN

The selection of components of adequate size for any structure involves two stages:

 (i) the *analysis* of the structure to determine the internal forces, shears and moments present at different points;
(ii) the *detailed design* of each component to cope adequately with these forces, shears and moments.

In simple structures, such as simply supported beams, analysis is a straightforward process, but in more complex frames, trusses, domes, etc., the analysis can be involved and time-consuming. Once the analysis has been completed, however, the detailed design always follows the same format: the selection of a cross-section of sufficient strength, stability and stiffness so that the internal forces revealed by the analysis will not lead to failure.

12.2   DESIGN OF A SIMPLE BEAM

Consider the design of a simple timber beam. We are concerned with *strength*: neither the *bending* stresses (tension or compression) nor the *shear* stresses must be allowed to be so large that the beam becomes unsafe. We are also concerned with *stiffness*: it is important that the beam should not deflect so much that associated elements of construction are damaged. Thirdly, we are concerned with *stability:* our entire structure, and also the elements of which it is composed, must remain stable under any likely pattern of loading. Strength, stiffness and stability — these are the attributes that we seek to build into structural elements like beams when we are undertaking their detailed design.

Timber has had an extensive use over the ages as a structural material. Because it is an extremely variable material, it is usually designed to have a very large margin

EXAMPLE 12.2

Check that the beam proposed in the previous example is adequate in shear and deflection.

Suggested beam size = 75 × 250 mm

Check shear

$$f_{s.max} = \frac{3}{2} \frac{V_{max}}{bd}$$

$$= \frac{3 \times 8 \times 10^3}{2 \times 75 \times 250} = 0.64 \text{ MPa}$$

$$f_{s.max} < 1.0 \text{ MPa}$$

$$\therefore 75 \times 250 \text{ OK in shear}$$

Check deflection

$$I = \frac{bd^3}{12} = \frac{75 \times 250^3}{12} = 97.5 \times 10^6 \text{ mm}^4$$

$$E = 10.5 \times 10^3 \text{ MPa}; \ C = \frac{5}{384}$$

$$\therefore \text{Deflection} = \frac{5}{384} \frac{W.l^3}{E.I}$$

$$= \frac{5 \times 16 \times 10^3 \times 4^3 \times 10^9}{384 \times 10.5 \times 10^3 \times 97.5 \times 10^6}$$

$$= \frac{80 \times 64 \times 10^3}{384 \times 10.5 \times 97.5} = 13.0 \text{ mm}$$

Allowable deflection:

$$= \frac{\text{span}}{300} = \frac{4000}{300}$$

$$= 13.3 \text{ mm}$$

$$\therefore 75 \times 250 \text{ OK in deflection}$$

$$\therefore \text{Use } 75 \times 250 \text{ hardwood beam}$$

before failure. We will limit the applied stresses in our timber beam to the values shown in Table 12.1, to ensure that the beam has adequate *strength* in bending and shear, under the working loads we will specify.

TABLE 12.1
Some Stress Limits (MPa) for Timber Beams under Working Loads

	Softwood	Hardwood
Bending (F_b)	5.5	11.0
Shear (F_s)	0.6	1.0
Modulus of Elasticity (E)	6.9×10^3	10.5×10^3

The *deflection* of a timber beam is calculated by the methods given in Chapter 11 using the value for E given in Table 12.1. The amount of deflection that can be tolerated in a beam depends upon its application. Some caution is needed because creep causes deflection to increase with time. The maximum allowable deflection is usually stated as a fraction of the span length. We will assume that the maximum deflection (short-term, using the tabulated E values) should not exceed 1/300 of the span:

$$\text{i.e. Maximum deflection} = \frac{\text{span}}{300}$$

By this criterion, a 6 m beam would be unacceptable if its maximum deflection exceeded

$$\frac{6000}{300} = 20.0 \text{ mm}$$

Lateral instability of a beam occurs if a beam is too narrow for its length and/or depth. Under these circumstances, the compression flange of the beam may be displaced sideways under quite small vertical loads.

In general terms, isolated timber beams should not have a depth greater than about three times the breadth, but beams acting as part of a connected system (e.g. floor joists, ceiling joists, rafters) may have a depth five or six times the breadth, provided that the beams are adequately attached to each other by bracing (e.g. blocking pieces).

12.3   DESIGN STEPS

Usually, timber beams are designed first for bending strength, and then checked for shear strength and deflection. The first step is then to select a beam cross-section for which the actual bending stress is less than the stress limit for the chosen

EXAMPLE 12.3

A floor is to be supported on hardwood timber joists that are spaced at 0.6 m centres and span 5.0 m. The joists support flooring 20 mm thick and an applied live load of 1.5 kN/m².

Design a typical joist 50 mm wide.

Data

Assume density of timber = 800 kg/m³
Bending Stress Limit   = 11.0 MPa
Shear Stress Limit   = 1.0 MPa
Young's Modulus   = 10.5 × 10³ MPa

The first step is to determine the loads for which the beam is to be designed. Each joist will support an area of floor 5.0 m long and 0.6 m wide, i.e. 3.0 m². The loading will be uniformly distributed to the joist. The mass of the flooring is determined from the known volume and density, and converted to weight.

Live load   = 3.0 × 1.5   = 4.5 kN
Dead load flooring = 3.0 × 0.02 × 800 × 10 = 0.48 kN
Dead load joist — assume: = 0.4 kN

∴ Total U.D. load = 5.4 kN approximately

$$BM_{max} = \frac{W.L}{8} = \frac{5.4 \times 5}{8} = 3.38 \text{ kNm; } SF_{max} = 2.7 \text{ kN}$$

$$\text{Required } Z = \frac{3.38 \times 10^6}{11.0} = 0.31 \times 10^6 \text{ mm}^3$$

$$\therefore d = \left[\frac{6Z}{b}\right]^{\frac{1}{2}} = \left[\frac{6 \times 0.31 \times 10^6}{50}\right]^{\frac{1}{2}} = 192 \text{ mm}$$

∴ 50 × 200 OK in bending

Check shear

$$f_{s.max} = \frac{3}{2}\frac{V_{max}}{bd} = \frac{3 \times 2.7 \times 10^3}{2 \times 50 \times 200} = 0.41 \text{ MPa}$$

∴ 50 × 200 OK in shear

type of timber. A check is then made to ensure that, for the beam so chosen, the shear and deflection are acceptable.

We will use lower case letters for the actual stress present in a beam, and upper case letters for the relevant stress limit. Thus:

f_b = actual bending stress for a particular beam

F_b = bending stress limit for the particular type of timber.

To design for bending stress, we make use of the section modulus Z. We know that the actual extreme fibre stress in a beam is given by:

$$f_b = \frac{M}{Z}$$

If we limit the maximum bending stress in the beam to a stress limit F_b, as in Table 12.1, then the maximum bending moment for that particular beam must be limited to:

$$M = F_b Z.$$

The product $F_b Z$ is sometimes called the *Moment Capacity* of a beam. We should put 'at working loads' after 'Moment Capacity' to avoid any confusion with 'ultimate moment capacity', which occurs at failure, as explained in paragraph 9.6.

Consequently, we can select *the minimum Z that will be required* for a particular beam, provided we know the maximum bending moment that will be applied and the stress limit at working loads for the chosen timber.

$$\text{Required } Z = \frac{\text{BM max}}{F_b}$$

Any beam having a cross-section with this (or a greater) value of Z will be safe in bending. The choice of the particular cross-section is then a compromise between the economy obtained from deep, narrow beams and the requirements for lateral stability. The lateral stability of a timber beam should be considered in conjunction with the cross-section needed for bending strength.

It must be remembered that the self-weight of the beam constitutes a load that must be carried in addition to the loads that are separately applied. Hence it is necessary to estimate the self-weight of the beam at the start of the analysis of the beam. A very precise estimate is not usually necessary for timber used in domestic construction; slender, short beams will weigh as little as 0.2 kN, and sturdy, long beams may weigh up to about 3 kN. It is usually sufficiently accurate to pick a reasonable value within this range.

The procedure for selecting a cross-section to satisfy bending strength and lateral stability is illustrated in Example 12.1.

EXAMPLE 12.3   (CONTINUED)

Check deflection

$$I = \frac{bd^3}{12}$$

$$= \frac{50 \times 200^3}{12}$$

$$= 33.3 \times 10^6 \text{ mm}^4$$

$$\therefore \text{Deflection} = \frac{5 \times 5.4 \times 10^3 \times 5^3 \times 10^9}{384 \times 10.5 \times 10^3 \times 33.3 \times 10^6}$$

$$= 25.1 \text{ mm}$$

$$\text{Allowable deflection} = \frac{5 \times 10^3}{300}$$

$$= 16.7 \text{ mm}$$

$\therefore$ 50 × 200 not OK in deflection

Try 50 × 250:

$$I = \frac{50 \times 250^3}{12}$$

$$= 65.1 \times 10^6 \text{ mm}^4$$

$$\therefore \text{Deflection} = \frac{33.3}{65.1} \times 25.1$$

$$= 12.9 \text{ mm}$$

$\therefore$ 50 × 250 OK in deflection

$\therefore$ Use 50 × 250 beam

Having selected a possible cross-section that is adequate in bending and stability, it is necessary to check this size for shear. The actual maximum shear stress in a rectangular beam is:

$$f_{s(max)} = \frac{3}{2}\frac{V}{bd}.$$

Provided this predicted shear stress is less than the shear stress limit, F_s, the beam will have adequate strength in shear. If the predicted stress is larger than the stress limit, a broader beam should be chosen and checked.

The predicted deflection of the beam should then be calculated, and compared with the permissible deflection. If the deflection is larger than permitted, a deeper cross-section should be chosen. The new predicted deflection can be calculated by proportion.

Example 12.2 shows the method.

12.4 ULTIMATE STRENGTH DESIGN

The *stress limits* in paragraph 12.3 are effectively *permissible stresses* as described in paragraph 2.5. They assume that once any tiny portion of a beam's cross-section reaches that limit, the beam reaches maximum permissible working load. Paragraphs 2.5, 9.6 and 10.5 refer to a more realistic approach called Ultimate Strength Design, which is based on the real, observed behaviour of materials at high stresses.

To use the ultimate (failure) capacity of a beam cross-section in bending or in shear, however, we must first apply a *Load Factor* to the working loads, to turn them into *ultimate loads*. These are loads that, if applied, could actually fail the structure. Commonly used load factors are 1.25 for dead load and 1.5 for live load. The live load factor is higher because it is less easy to estimate with certainty.

For extra safety, a *Capacity Reduction Factor* (less than 1.0) is applied to the ultimate moment capacity (and shear capacity) of a cross-section, typically to represent uncertainties about quality of materials, of construction and of calculations.

While 'permissible stress design' prevailed in the early days of structural engineering, most codes in most countries have now adopted Ultimate Strength Design in most common structural materials because it better reflects real material behaviour. It is obviously important in the ultimate strength method to remember to apply load factors and capacity reduction factors at the right step. Otherwise, the basic approach to structural design remains as outlined in paragraph 12.1.

The Permissible Stress Design method, based on linear elastic stress theory, is still a valuable introduction to element design and is still commonly used for preliminary design purposes in all kinds of situations.

WORKSHEET 12.1

12.1 The sketch shows the structural roof timbers for a small building. The roof is to be supported by two walls 5 m apart, and consists mainly of softwood rafters A, 0.8 m apart. However, the end rafter B is required to support a beam C (which in turn carries another part of the roof not shown in the sketch). Beam B will be hardwood.

(a) Design a typical rafter A. The rafters support roof-sheeting weighing 4 kg/m² and a straw-board ceiling weighing 12 kg/m².

(b) The rafters will *also* have to support the weight of workmen who may need to do maintenance on the roof. If this maintenance load is 0.25 kN/m², what size rafters would be required for bending strength?

(c) Design beam B. Assume that it carries a uniformly distributed load (inclusive of its own weight) of 1.2 kN/m, and a concentrated load of 4 kN applied at midspan from beam C.

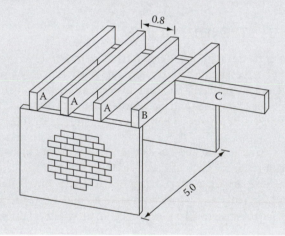

APPLICATION: TIMBER BEAMS, NOTTINGHAM UNIVERSITY
JUBILEE CAMPUS (SEE PHOTOGRAPHS OVER PAGE)

Minimising energy use and CO_2 production were key drivers for the design of the new Jubilee Campus of Nottingham University, UK, including the School of Management and Finance building. The use of glued-laminated ('glulam') timber beams for the atrium roof structure contributed to this strategy.

Glulam beams involve several production processes, from kiln drying, stress grading, and finger jointing, to laying up in laminate presses, end joint preparation, surface finishing, and protective coating. Superior timber grade, and finger joint proof testing, are often used for the outermost laminates where bending stresses are highest. The inter-laminate glue joint experiences longitudinal shear stress (see Chapter 10), though usually at a low level. Pre-camber against dead load deflection can be fabricated into the beams.

Project: Jubilee Campus, University of Nottingham, UK; Architect: Michael Hopkins and Partners; Structural Engineer: Arup; Photos: Paul McMullin; Arup

APPLICATION: TIMBER BEAMS, DRUK WHITE LOTUS
SCHOOL, LADAKH (SEE PHOTOGRAPHS OVER PAGE)

The Druk White Lotus School project in Ladakh was initiated by the Drukpa Trust under the patronage of His Holiness the Dalai Lama, and will eventually cater for 750 mixed pupils from nursery to 18 years. Planning, design and construction are being carried out with great respect for the local ecology, and for the sustainable management of water, waste, energy and internal environmental control.

The roof for the first classroom buildings comprises a mud coating laid on a mat of hardy grass from the local Indus valley, supported in turn by a layer of willow lath (running left to right in the photo), sourced from monastery plantations nearby. There are then poplar joists of minimum 150 mm diameter at 450 mm centres, spanning onto pine main frames (Butu and Kale), which were estimated to be C16 grade for structural design to Eurocode 5. Steel plates are slotted and bolted into the beam-to-column joints of the timber frames for resistance to lateral earthquake loads in this highly seismic region.

Architect: Arup Associates; Engineer: Arup Associates and Arup; Photos: Arup Associates (Caroline Sohie, Roland Reinardy)

Answers to Worksheets

SHEET 1.1

1.1 Component in direction $AB = 8.66$ kN, $CD = 7.07$ kN, $EF = 2.59$ kN.
1.2 (a) No.   (b) 6.06 kN @ 33° to horizontal.
1.3 (a) 40 kNm.   (b) 2 m.   (c) 20 kN.
     (d) 20 kN @ 30° to horizontal; $V = 10$ kN, $H = 17.3$ kN.
1.4 (a) 800 N.   (b) 107.2 N.
     (c) 807 N @ 7° to vertical.

SHEET 1.2

1.1 (a) $H_A = 3$ kN, $V_A = 2$ kN, $M_A = 14.5$ kNm.
     (b) $H_A = 4$ kN, $V_A = 9.3$ kN, $V_B = 15$ kN.
     (c) $V_A = 27.3$ kN, $H_D = 4$ kN, $V_D = 83.3$ kN.
1.2 (a) $F_{AD} = 5.83$ kN, $F_B = 5.83$ kN.
     (b) $H_C = 0$, $V_C = 6$ kN, $M_C = 15$ kNm.
1.3 $F_{AB} = 1.0$ kN, $F_{BC} = 1.225$ kN, $F_{BD} = 1.367$ kN.

SHEET 2.1

2.1 (a) $H_D = 0$, $H_C = 2$.   (b) 2 kN (strut).
     (c) $V_C = 2$ kN.   (d) $BC = 2.83$ kN (tie), $AC = 0$.
2.2 (a) 7.07 kN,   90 MPa.
     (b) 7.07 kN,   45 MPa.
     (c) $V = 5$ kN, $H = 5$ kN, $M = 10$ kNm.
2.3 (a) 95.9 kN.   (b) 38.4 kN.   (c) 9.6 kN.
     (d) 614.1 kN.
2.4 $AB = BC = 13.2$ kN, $BD = 1.3$ kN, $BE = 0$.

SHEET 3.1

3.1 (a) 1.0 kPa.   (b) 0.6 kN/m.
     (c) AB 2.72 kN/m; CD 0.22 kN/m + 8.16 kN.

SHEET 4.1

4.1 (a) 4.91 kN @ 48°.   (b) 0.152 m³.
4.2 800 N (tension); 693 N (compression).
4.3 300 kN.
4.4 400 N and 500 N @ 36.9°.

SHEET 5.1

Figure 1:   B1 + 3, A1 – 2.65, 12 + 0.88, C2 + 2.5, A3 – 2.65,
            23 – 0.87, 45 + 0.88, D4 + 2.5, A5 – 2.65, 34 – 0.87, E5 + 3.
Figure 2:   B1 – 7.04, A1 + 4.98, A2 + 4.98, 12 0, 23 + 4.22, C3 – 8,
            D4 – 8, 34 – 1.5, 45 – 2.11, A5 + 9.47, A6 + 9.47, 56 0.
Figure 3:   D5 + 2.83, C5 – 2, 45 – 2, D4 + 2, 34 + 5.66, B3 – 6,
            23 – 4, D2 + 6, 12 + 11.31, A1 – 14, F10. (tension –ve).

SHEET 7.1

7.1   (a)   Max. S.F. = –7.33 kN near D;   Max. B.M. = +4.67 kNm @ B.
      (b)   Max. S.F. = –9.2 kN near C;   Max. B.M. = –5.5 kNm @ C.
      (c)   Max. S.F. = +12 kN near B;   Max. B.M. = –20 kNm @ B.
      (d)   Max. S.F. = +18 kN near B;   Max. B.M. = –16 kNm @ D.
7.2   1.15 m from D, and 0.76 m from B.

SHEET 8.1

8.1   (a)   x = 1.18 m   (b)   No.
8.2   (a)   y = 175 mm.
      (b) $I_{yy} = 114.6 \times 10^6$, $I_{xx} = 130.2 \times 10^6$,
            $Z_{yy} = 0.76 \times 10^6$, $Z_{xx\,max} = 1.74 \times 10^6$, $Z_{xx\,min} = 0.74 \times 10^6$.
8.3   (a)   y = 613 mm.
      (b) $I_{xx} = 23100 \times 10^6$, $Z_{xx\,max} = 59.7 \times 10^6$, $Z_{xx\,min} = 37.7 \times 10^6$.

SHEET 9.1

9.1   At AA: 4.0, 3.0, 2.0 MPa compression, 2.0 MPa tension.
      At BB: 2.0, 1.5, 1.0 MPa compression, 1.0 MPa tension.
9.2   At B: 4.88 and 11.49 MPa.
      At C: 8.62 and 20.27 MPa.

SHEET 10.1

10.1   12.0 MPa;   0.3 MPa.
10.2   24 MPa. The provision of adequate shear strength enables the first beam
       to carry *twice the load* for a given level of bending stress.
10.3   (a)   0.71 MPa.   (b)   0.72 MPa.   (c)   1.16 MPa.

SHEET 11.1

11.1   (a)   3.0 mm.   (b)   4.8 mm.
11.2   12.0 mm;   12.7 mm.
11.3   10.1 + 11.6 = 21.7 mm.

SHEET 12.1

12.1 (a) 125×50 satisfactory in bending and shear; need 175×50 for deflection.

      (b) 175×50.

      (c) 300×75.

Appendix
A First Encounter with Statics

INTRODUCTION

Students come to the study of architecture or building equipped with a variety of educational experiences. Some have studied engineering science or physics during their secondary studies, and are familiar with most of the concepts and much of the detail presented in this book.

Chapter 1 of this book — Forces, Moments and Equilibrium — was designed for people who have met these concepts before. The chapter merely gathers together these basic and familiar concepts as a refresher course before moving onto the material in the chapters that follow.

If you have *not* previously or recently encountered the concepts and procedures reviewed in Chapter 1, then this Appendix is for you.

1 WHAT YOU NEED TO KNOW ABOUT TRIGONOMETRY

1.1 In trigonometry we are concerned with lines and the angles between them. We measure angles in degrees: there are 90 degrees (90°) in a right angle, 180° in a straight line and 360° in a full circle (Figure A.1).

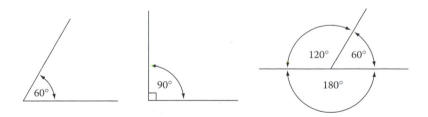

FIGURE A.1   Trigonometry is about angles.

1.2 Start with the right-angled triangle ABC is which the angle at B is a right angle, 90° (Figure A.2). In any triangle, the sum of the internal angles is 180°, so that in a right-angled one the sum of the angles at A and C must be 90°; this is usually expressed in the following way:

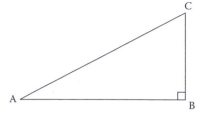

$$A + B + C = 180°$$

$$A + C = 90°$$

FIGURE A.2 Trigonometrical ratios of the right-angled triangle.

1.3 The ratio of the side AB to the side AC is called the *cosine* of the angle A, written as *cos* A. The side AC (the longest side of the right-angled triangle) is called the *hypotenuse*, and AB can be called the *adjacent side* (because it adjoins the angle A).

So we could write: $\cos A = \dfrac{\text{adjacent side}}{\text{hypotenuse}} = \dfrac{AB}{AC}$

The other ratios for the angle A involve the side BC, called the *opposite side*. They are called the *sine* and the *tangent* of A (written *sin A* and *tan A* respectively):

$$\sin A = \frac{\text{opposite side}}{\text{hypotenuse}} = \frac{BC}{AC} \qquad\qquad \tan A = \frac{\text{opposite side}}{\text{adjacent side}} = \frac{BC}{AB}$$

For the angle C we have:

$$\cos C = \frac{\text{adjacent}}{\text{hypotenuse}} = \frac{BC}{AC} \qquad\qquad \sin C = \frac{\text{opposite}}{\text{hypotenuse}} = \frac{AB}{AC}$$

1.4 We see that in a right-angled triangle, the sine of one angle is exactly equal to the cosine of the other (i.e. sin A = cos C).

1.5 Sometimes we need to work with triangles that do not contain a right-angle, such as the triangle ABC (Figure A.3). Letters a, b and c represent the sides opposite the angles A, B and C respectively.

There are two 'rules' that geometricians find useful here: the sine rule and the cosine rule:

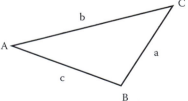

$$\frac{a}{\sin A} = \frac{b}{\sin B} = \frac{c}{\sin C} \quad \text{Sine rule}$$

$$c^2 = a^2 + b^2 - 2.a.b.\cos C \quad \text{Cosine rule}$$

FIGURE A.3 The sine and cosine rules can be used for any triangle.

Exercises 1

1.1 For the following right-angled triangles, find the angles or sides marked with a question mark:

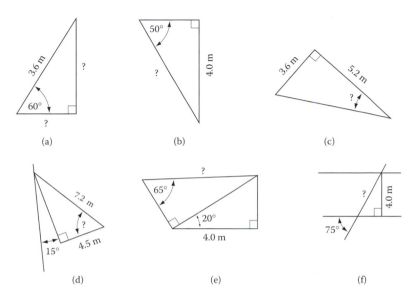

FIGURE E.1

1.2 For the following triangles, find the angles or sides marked with a question mark:

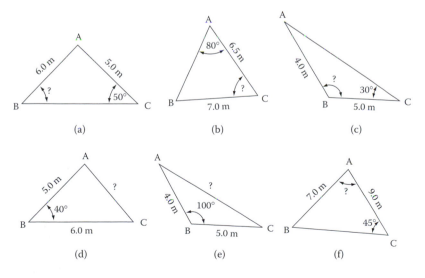

FIGURE E.2

2 EXTERNAL FORCES

2.1 'Force' is the name given to a push or a pull applied to an object.

A force must have an object on which to act. Unless there is something that can be pushed or pulled there is no force. The object doesn't have to move in response to the force (although it may) — it just has to be there. The easiest way to imagine forces is to consider that you are pushing the object with your finger, or pulling it with a piece of string (Figure A.4). You will then see that every force has three characteristics:

- sense and direction (is it a pull or a push, and in what direction?);
- magnitude (how big is the pull or push?); and
- location (at what point is it applied?).

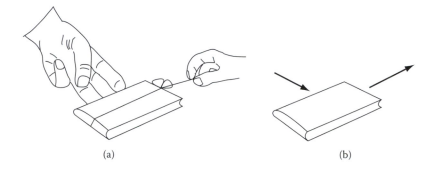

(a)   (b)

FIGURE A.4   A force is a push or a pull applied to a body.

2.2 We will use an arrow to represent a force. If the arrow is directed towards the object, it represents a push, as in Figure A.4. A push from the left will have the same effect on the object as a pull from the right, and we can interchange the two.

The most commonly encountered force is the gravitational force with which the earth pulls an object. We call this the *weight* of the object. The basic unit of force is the Newton (N). The weight of an apple is about one Newton. In buildings, we deal in thousands of Newtons (kilonewtons, kN), and the weight of a heavy person would be about one kilonewton.

3 THE RESULTANT OF TWO FORCES

3.1 When we talk about the resultant of two forces, we are speaking of *a single imaginary force that has the same effect on the object as the totality of the two real forces that actually act on the object.* Suppose we have a large rock R that we wish to move to a new position on a site (Figure A.5a). We attach two tractors (A and B) to it, and we start to pull, to drag the rock in the direction RC. The effect on the rock would have been exactly the same if we had used a single tractor pulling at C. We can imagine a single force acting in the direction RC which has the same effect as the two real forces acting in the directions RA and RB — this force is the *resultant* of the other two forces.

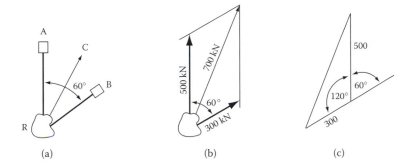

FIGURE A.5   The forces applied by the two tractors at A and B could be replaced by an imaginary resultant force at C; the size and direction of this resultant can be found by drawing a parallelogram of forces.

3.2 The direction and size of the resultant depend upon the directions and sizes of the two applied forces. Clearly, if the tractor at B pulled a little less strongly, the resultant would be a little smaller and would be directed a little closer to A. If B stopped pulling altogether, the resultant would be identical to the force produced by tractor A. If both A and B pulled twice as hard in their present directions, the resultant force would be twice as large, but still in its present direction.

3.3 We can find the resultant of a pair of forces by making a scale drawing, which we call a *Parallelogram of Forces*.

Each force is represented by an arrow, which has a length proportional to the size of the force. Suppose that the tractor A is pulling north on the rock with a force of 500 kN (Figure A.5b). To construct the parallelogram of forces, draw the rock and then draw a northerly pointing arrow 50 mm long. This arrow represents the force to a scale of 1 mm = 10 kN. If the other tractor is pulling at an angle of 60° with a force of 300 kN, represent that force with an arrow 30 mm long at 60°. To find the resultant force, complete the parallelogram graphically and then draw in the diagonal arrow. The length and direction of this arrow represent the size and direction of the resultant force.

3.4 Rather than using graphical (drawing) methods, it is often more convenient to calculate the sizes and angles of the parallelogram using trigonometry. If you look at one of the two triangles that make up the parallelogram, you will see that use of the cosine rule will immediately give the size of the resultant (700 kN), and the sine rule will give its direction (21.8°).

All problems in statics can be solved by either of two methods — graphical (by drawing to scale) or analytical (by calculation).

3.5 Often you can do a rough sketch, and quickly get an estimate of the size and direction of the resultant of two forces, using the parallelogram of forces method. Figure A.6 shows some examples; note that in Figure A.6c we have transposed a push P from the right into a pull P towards the left.

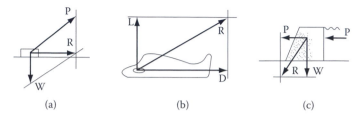

(a)   (b)   (c)

FIGURE A.6 Examples of resultant forces: (a) Pulling a heavy object across the floor; (b) Lift and drag on an aeroplane wing; (c) Pressure on the back of a heavy retaining wall.

3.6 A particular application of the parallelogram method is when we are dealing with two forces that happen to be at right angles — for example, the weight of a building and the push of the wind that may be acting on it (Figure A.7). The parallelogram for a pair of perpendicular forces is a rectangle, making the geometry particularly easy to solve.

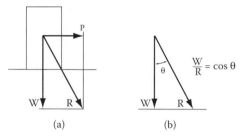

(a)   (b)

FIGURE A.7 The resultant of a pair of perpendicular forces.

4 THE COMPONENTS OF A FORCE

4.1 Figure A.7 showed two forces that were perpendicular to each other, and the single imaginary force — their resultant — that has the same effect. We could read this same diagram another way: as a single *applied* force and a pair of perpendicular *imaginary* forces that produce the same effect. Figure A.8 shows this more clearly. F is an inclined force applied to an object, and C1 and C2 are a pair of imaginary forces that produce the same effect.

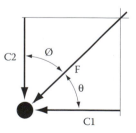

FIGURE A.8 C2 and C1 are the vertical and horizontal components of the force F.

$$\frac{C1}{F} = \cos\theta \quad \text{and} \quad \frac{C2}{F} = \cos\varnothing$$

Why would we want to replace one real force with a pair of imaginary ones?

4.2 Often, we are interested to know what effect a force might have in a direction other than its own. For example, Figure A.9 shows a mop being pushed across a floor. The applied force is straight along the handle, at an angle of A to the floor. The applied force is in the direction of the mop-handle, but what effect does this force have in the direction of the floor?

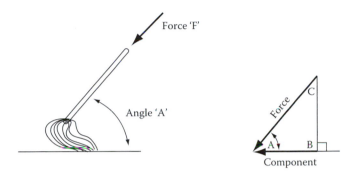

FIGURE A.9 A force F in the direction AC can produce an effect in the direction AB; we say that the force F produces a component force in the direction AB.

4.3 *The effect (in a particular direction AB) of a Force F (which is acting in some other direction AC) is called the component of the force in direction AB.*

The component and the original force form a cosine relationship:

$$\cos A = \frac{\text{component AB}}{\text{force AC}}$$

This can be rewritten as: component = force × cos A

This is an exceptionally useful relationship that you will use over and over again: *The component of a force is equal to the product of the force itself and the cosine of the included angle.*

4.4 When you wish to find *both* perpendicular components of a force, it is simpler to use the cosine relationship for one component and the sine of the *same angle* for the other.

Flying buttresses were a feature of Gothic cathedrals. They transmitted an inclined force from the cathedral roof to a freestanding pier (Figure A.10). The inclined force

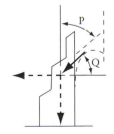

FIGURE A.10 The vertical and horizontal components of the force in a flying buttress.

has two effects: a horizontal component (force × cosine of horizontal angle, Q) which tends to push the pier over, and a vertical component (force × cosine of vertical angle, P, or force × sine of horizontal angle) which tends to push the pier straight down into the ground.

EXERCISES 2

2.1 Draw parallelograms of force to find the resultants in the following examples, then confirm your answers using trigonometry.

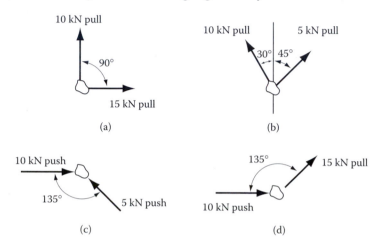

FIGURE E.3

2.2 Find the value of the component in direction a–a for each of the following examples.

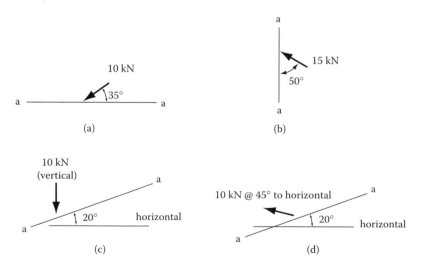

FIGURE E.4

2.3 What is the resultant of the three forces shown in Figure E.5a?

2.4 In Figure E.5b, tractor A is pulling north with a force of 300 kN. In what direction should tractor B (500 kN) pull so as to move the rock in a direction 30° east of north?

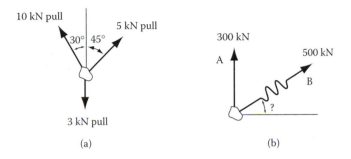

FIGURE E.5

5 EQUILIBRIUM OF FORCES

5.1 We say that an object is in equilibrium if it is in a state of rest or if it is moving at constant velocity.

An object in equilibrium has *no resultant external force acting on it.* A book lying on a desk has two forces acting on it: the earth is pulling it downwards and the desk is pushing it upwards *with exactly the same amount of force.* The downward force and the upward force are exactly equal to one another; they cancel each other out, and do not produce any *resultant* force in a vertical direction. If we add together all the vertical forces acting on the book, the sum will be zero. However, if the upward force is increased it will be larger than the downward one, there *will* be a

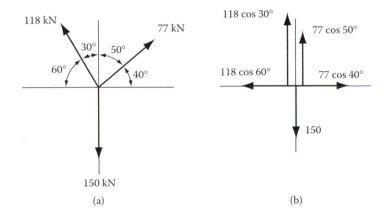

FIGURE A.11 Inclined forces can be replaced by their vertical and horizontal components in order to see whether the forces are in equilibrium.

resultant (upward) force, the sum of all vertical forces will no longer be zero, and the book will accelerate upwards.

There may be a large number of forces acting on an object, some vertical, some horizontal and some inclined (Figure A.11a). We can take all of the inclined forces and replace them with their vertical and horizontal components, so that the system of forces now consists of vertical and horizontal forces only (Figure A.11b). If this object is to be in equilibrium in a vertical direction, the sum of all vertical force components must be zero. We sometimes express this shorthand as *sigma V equals zero* or *sigma Fy equals zero*; the Greek letter sigma (Σ) stands for 'the sum of'.

$$\Sigma V = 0 \quad \text{or} \quad \Sigma Fy = 0$$

If the object is in equilibrium in a horizontal direction, the sum of all the horizontal force components will be zero:

$$\Sigma H = 0 \quad \text{or} \quad \Sigma Fx = 0$$

5.2 Figure A.11 shows the forces on a heavy girder weighing 150 kN being lifted into position by two cranes, which supply forces of 118 kN at 60° and 77 kN at 40° to the horizontal. If we replace these applied forces by their vertical and horizontal components, we see that the horizontal components are 118.cos 60° (i.e. 59 kN) acting to the left and 77.cos 40° (59 kN) to the right. The girder *is* in equilibrium in a horizontal direction. The vertical components are 102.2 kN and 49.5 kN upwards and 150 kN downwards; there is a resultant upward force of 1.7 kN, so the girder *is not* in vertical equilibrium and will accelerate upwards.

The process of summing the component forces is more straightforward if we give signs to the forces depending upon their sense. For vertical forces, the usual convention is that upward forces are considered to be positive and downward forces negative. For horizontal forces, those acting to the right are positive and those to the left are negative. Thus, in the previous example, we could eliminate all the words and write:

$$\begin{aligned}
\Sigma Fx &= -188 \cos 60° + 77 \cos 40° \\
&= -59 + 59 \\
&= 0 \\
\Sigma Fy &= +118 \cos 30° + 77 \cos 50° - 150 \\
&= +102.2 + 49.5 - 150.0 \\
&= +1.7 \text{ kN}
\end{aligned}$$

5.3 Because statics deals with forces acting on objects that are at rest, the concept of equilibrium is fundamental to all we do. The structures we use in buildings are in equilibrium. Typically, every part of every building structure is in equilibrium.

5.4 So far we have been considering only force systems in which all the forces lie in the one plane and all pass through the one point. These are called *co-planar, concurrent* systems of forces. For these, we can make use of the two *equations of equilibrium:* $\Sigma V = 0$ and $\Sigma H = 0$. We can use these two equations to find two unknown quantities — two unknown force magnitudes, or two unknown directions, or an unknown magnitude and an unknown direction.

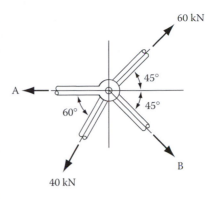

FIGURE A.12 Using two equations of equilibrium to find unknown forces in a bolted joint in a structural framework.

5.5 Figure A.12 shows a joint in the structural framework of a building, where a number of steel bars are held together with a single bolt. The forces at A and B are unknown; what are they? There are only two unknowns — the magnitudes of the two forces — and we have two equations of equilibrium. We consider equilibrium in two directions, and set up two equations. Considering the vertical direction, we have:

$$\Sigma V = 0$$

i.e. $+ 60 \cos 45° - 40 \cos 30° - B \cos 45° = 0$, which gives B = 11.0 kN.

(The force A is horizontal and hence has no vertical component.) Similarly, we could set up an equation for horizontal equilibrium, which gives A = 30.2 kN.

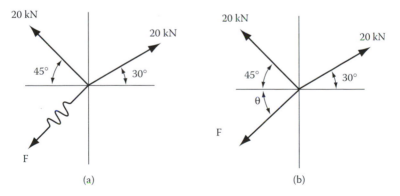

FIGURE A.13 How to deal with a force for which both magnitude and direction are unknown.

5.6 Figure A.13a shows three forces acting on an object; the wriggly arrow is used to represent a force acting in an unknown direction. Again we have

two unknowns — this time, the magnitude and direction of the force F —
so we can use the two equations of equilibrium. If the angle between F and
the horizontal is theta (Figure A.13b), we can write down the equation for
vertical forces:

$$\Sigma V = 0$$
$$+ \ 20 \cos 45° + 20 \cos 60° - F \cos (90° - \theta) = 0$$

Replacing $\cos(90° - \theta)$ by $\sin \theta$ (see paragraph 1.4) and applying the horizontal equilibrium equation as well, gives F = 24.34 kN, at 82.5° to the
horizontal.

6 MOMENT OF A FORCE

6.1 When a force is applied to an object, it may cause it to rotate about some
point. When we pull on the end of a spanner, when we sit on a see-saw,
or when we push down on a bicycle pedal we are using a force to cause a
rotation (Figure A.14). We call the rotational effect of a force the *moment
of the force about the point*, or, more simply, the *moment*. We measure
the moment by the product of the force and the 'lever arm', which is the
perpendicular distance from the point to the force:

Moment = force × lever arm

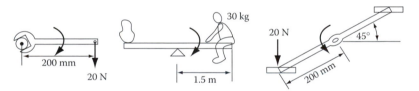

FIGURE A.14 Example of forces that produce moments causing rotation about
a point.

Being the product of a force and a distance, moments have the units
of *newton metres*. In building construction, the preferred unit is the
kilonewton metre, kNm. If the moment tends to cause a clockwise rotation of the object, we give the moment a positive sign; if the rotation is
anticlockwise, the moment is negative.

A curved arrow as in Figure A.14 represents the moment of a force. An
arrow representing a positive moment points in a clockwise direction.

6.2 Figure A.15 shows some examples of forces acting at various different
lever arm distances from point A. You may need to extend the line of
the force arrow in some cases to discover the *perpendicular* distance (the
lever arm) from A.

6.3 In many examples there will be an obvious point about which an object
will rotate. But we can work out the moment of a force about *any point* at

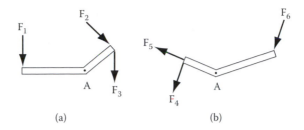

(a) (b)

FIGURE A.15 Practice drawing the lever arm from A to each of the forces shown.

all — the point need not be an axle or fulcrum; it could just as easily be an edge or corner, or the place where the object touches some other object. The important thing about the moment of a force is that it *tends* to cause a rotation; it does not matter whether the object is physically capable of rotating about that particular point. In statics, of course, we are concerned about objects that are in equilibrium and not moving, so it does not matter which point we choose. As long as the object remains in equilibrium, it will not in fact rotate about any point. Moment will always be the product of force and perpendicular distance from the point.

6.4 For an object to be in rotational equilibrium, it must have no resultant moment acting about any point. That is, the sum of the moments about any point must be zero.

$$\Sigma M \text{ (about any point)} = 0$$

7 CO-PLANAR FORCE SYSTEMS

7.1 In much of our studies of structure, we will only be concerned with two-dimensional structures, that is, structures that lie wholly within the one plane. Many everyday structures (beams, trusses, braced frames, rigid frames etc.) can be considered as two-dimensional. The majority of three-dimensional buildings are constructed of arrangements of two-dimensional structures.

7.2 If we are dealing with structures that are two-dimensional, then we will be dealing also with forces that lie only within one plane. The forces will be co-planar, and therefore we need only concern ourselves with the three equations of equilibrium that we have encountered so far:

$$\Sigma V = 0$$

$$\Sigma H = 0$$

$$\Sigma M = 0$$

We can use these three equations to find any three unknown characteristics of our system of co-planar forces — the location, the direction or the magnitude of any of the forces acting on our structure. If, in the system of forces, there are *more* than three unknown quantities, we cannot find them with these methods and the system is called *indeterminate*. In creating buildings, we make use of both determinate and indeterminate systems without discrimination, but in our use of statics here, we only analyse the determinate ones.

7.3 Figure A.16 shows a beam ABC which we require to be in equilibrium under the action of five forces, of which three have unknown magnitudes. Clearly, the forces are not concurrent. We should be able to use the three equilibrium

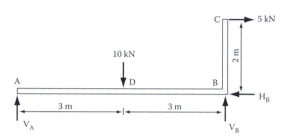

FIGURE A.16 Using all three equations of equilibrium to find three unknown forces.

equations to find the three unknown quantities. The equation for horizontal components tells us that H_B is 5 kN, but we can't learn anything immediately from the vertical force equation because there are *two* unknown vertical forces. If, however, we consider moments about the point A we have:

$$\Sigma M_A = 0$$

$$\text{i.e. } + 10 \times 3 - V_B \times 6 + 5 \times 2 = 0$$

from which we find that V_B is 6.67 kN. We could now consider vertical forces to find V_A or we could consider moments about any other point such as B or C to find that V_A has a value of 3.33 kN.

In Figure A.17 a beam AB is supported by a wall and by a cable CB. Again there are three unknowns: the force in the cable, and the direction and size of the force from the wall at A. By considering moments of forces about the point A, we find that the value of the force T is 10 kN. We then have two equations left, for vertical and horizontal components, and we have two unknowns. The unknown force is 10 kN and the unknown angle is 60°.

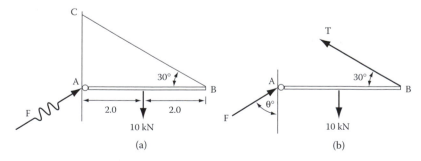

FIGURE A.17 Another use of the three equilibrium equations.

Although it is true that the sum of moments about *any* point is zero, some points will lead much more quickly to a solution. In Figure A.16, to find V_B, it is best to consider moments about A, because in the resulting equation there will be only one unknown quantity. Taking moments about some other point (D, for example), the equation will include *two* unknowns, V_A and V_B. When using the equation for moments, it is best to choose a point on the line of at least one of the unknown forces, to prevent that unknown from appearing in the equation.

EXERCISES 3

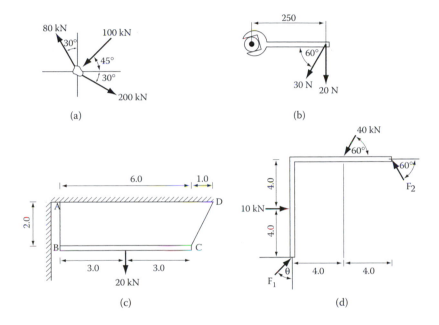

FIGURE E.6

3.1 Figure E.6a shows an object being acted upon by three forces. Is the object in equilibrium? If not, what is the magnitude and direction of the force that would need to be applied to restore equilibrium?

3.2 Figure E.6b shows a spanner being used to tighten a nut. What moments are produced on the nut by each of the forces?

3.3 Is the suspended beam shown in Figure E.6c in equilibrium? If not, suggest two ways of making it so, and try to work out the forces involved in each. If the structure were not stabilised in such a way, it would change its geometry until it reached equilibrium; what would the forces in the two suspension cables AB and CD be then?

3.4 The structure shown in Figure E.6d is in equilibrium. Calculate the forces F_1 and F_2 and the unknown angle θ.

8 REACTION FORCES

8.1 For convenience, we often divide the forces that act on a structure into two different groups: the *active* forces and the *reactive* ones. Active forces are usually those imposed by the environment: wind, weights of the object carried by the structure, human activities that the structure carries, and so on. Sometimes we call these active forces the *loads* that are applied to the structure. The reactive forces (often called the *reactions*) are the forces that are produced at the support points as a consequence of the active forces. If this book is resting on a table, and we consider the forces acting *on the book*, the downward weight of the book is an active force and the responsive upward force of the table is a reactive force. If you lean your elbows on the table, and we consider the equilibrium of the *table*, the weight of the book, the downward forces of your elbows and the weight of the table itself are all active forces, and the upward supporting forces at the bottom of the legs of the table are the reactions.

This subdivision into active and reactive forces is of course arbitrary, but convenient for analysis and discussion. Certainly, as far as the structure is concerned, there is no distinction between the two and the active forces and the reactions are treated the same way when we are using equations of equilibrium. The main distinction is that, when we first start to study a particular structure, the active forces (or loads) are usually known, and the reactions at the supports are unknown.

8.2 We need to be aware of the different types of reactive forces that different support devices can provide:

Roller: Supplies a *single unknown force*, in a direction perpendicular to the surface on which the roller rests; this force can be in either sense i.e. upwards or downwards).

Hinge: Supplies *two unknown component forces*, vertical and horizontal. Or we can consider the hinge as supplying a single unknown force inclined in an unknown direction. (See, for example, Figure A.17.)

Fixed support: Supplies *three unknowns*, an unknown vertical force, an unknown horizontal force and an unknown moment.

8.3 Most exercises or problems will include a diagram with support symbols representing one or other of these support types. It is best to draw a fresh diagram that removes the support symbols and replaces them with the *forces* that the type of support can apply. A hinged support, for example, would be replaced with an unknown vertical force and an unknown horizontal force. This new diagram is then called a Free-Body Diagram (F.B.D.). It shows the object being studied (e.g. a beam) and *all the forces that act on it.*

The F.B.D. does not show surrounding supports or walls or posts or hinges, or anything except the object and the forces that act on it. A carefully sketched F.B.D., roughly to scale, is of enormous help in problem-solving.

In drawing the Free-Body Diagram, we usually have to assume a sense for unknown forces and moments. Is this unknown reaction force up or down? Is that unknown moment at a fixed support clockwise or anticlockwise? Some texts suggest that you always assume one thing; for example, that all unknown moments are clockwise. It can be more instructive to look at the applied forces and try to deduce the sense of each unknown. Often it will be quite obvious, but sometimes you will have to make a guess. Then, when you have done the calculations, look at the sign of your answer. If the answer is positive your initial guess was correct: the force *does* have the sense that you expected. If the answer is negative, your guess was wrong: the force actually goes in the opposite direction to what you expected. Finally, draw a new diagram, with the forces in the right direction.

8.4 Figure A.18 shows a hinge-based cable-supported structure that carries an inclined 5 kN force. The first step in analysing such a structure is to draw the Free-Body Diagram, assuming senses for the unknown forces at the supports. We have three unknown reactions: F_A, V_B and H_B.

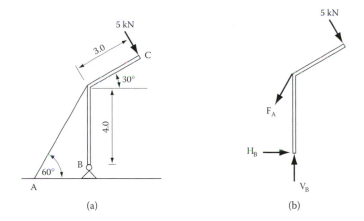

(a)   (b)

FIGURE A.18   Use of a Free-Body Diagram for a structure supported by a hinge and a cable.

Consider moments about point B (where two unknown forces intersect), to produce an equation containing only the unknown F_A. From this we find that F_A is +12.5 kN, so we know that our assumed sense was correct. It is then a straightforward matter to use the other two equations to find the other two unknowns (+15.16 kN and –3.75 kN). The minus sign tells us that the sense we assumed for H_B was incorrect; H_B in fact is directed towards the left.

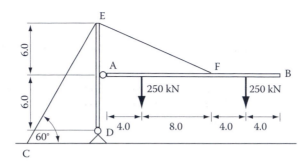

FIGURE A.19   An example of a structure that can be solved by drawing separate Free-Body Diagrams for each rigid element.

Figure A.19 shows a beam AFB supported by a hinge at A and by a cable at F. The hinge and cable are attached to a mast DAE, which is also supported by a hinge (at D) and a cable CE (at E). Structures such as this, which are composed of a number of components joined together with cables and hinges, are best broken up into F.B.D.s of their component substructures or parts.

Consider first the beam AFB (which has loads applied to it), and draw a F.B.D. for it. The F.B.D. will have three unknown forces — the tension in the cable EF and the reactions in the hinge — so can be solved with the three equilibrium equations. The force in the cable is 932 kN and the vertical and horizontal reactions at the hinge are 83 kN (upwards) and 833 kN (to the right). Now draw a F.B.D. for the next rigid component, the mast DAE. Forces will be applied to this mast via the cable EF and the hinge at A, and we know what these forces are. The forces exerted (by the beam) on the mast will be the same as those exerted (by the mast) on the beam, but they will have the opposite sense: the cable will pull upwards to the left on the beam at F, but downwards to the right on the mast at E. At the hinge, the vertical reaction on the beam was 83 kN upwards, so the force applied to the mast will be 83 kN downwards. In this way, you will find that the mast also has only three unknown forces on it, and once again you can apply the three equations of equilibrium to find them; they are 833 kN, 417 kN and 1222 kN.

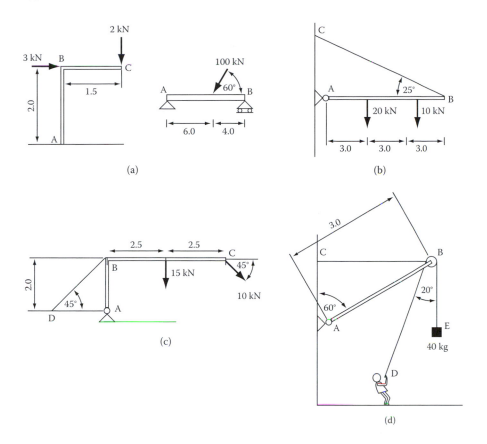

FIGURE E.7

EXERCISE 4

4.1 Calculate the reactions at the supports for the two structures shown in Figure E.7a.

4.2 The beam AB (Figure E.7b) is supported by a hinge at A and a cable at C. Calculate the reactions.

4.3 A bent beam (also called a *cranked* beam) ABC (Figure E.7c) supports two forces of 15 kN and 10 kN. What is the size of the tension in the supporting cable DB and what are the reactions at A?

4.4 A crane AB is attached to a building by a horizontal cable CB as shown in Figure E.7d. A pulley is attached to the crane at B, and a man is pulling on a rope at D to lift the 40 kg box at E. Find the force in the cable.

ANSWERS TO APPENDIX EXERCISES

EXERCISES 1

1.1 (a) 1.8; 3.12   (b) 5.22   (c) 34.7°
     (d) 51.3°   (e) 4.70   (f) 4.14

1.2 (a) 39.7°   (b) 33.9°   (c) 111.3°
     (d) 3.9   (e) 6.9   (f) 69.6°

EXERCISES 2

2.1 (a) 18.0 kN @ 33.7° to horizontal
     (b) 12.3 kN @ 6.9° to vertical
     (c) 7.4 kN @ 28.7° to horizontal
     (d) 23.2 kN @ 27.2° to horizontal

2.2 (a) 8.2 kN
     (b) 9.6 kN
     (c) 3.4 kN
     (d) 4.2 kN

EXERCISES 3

3.1 119.1 kN @ 58.3° from horizontal (downwards to left)

3.2 5.0 Nm, 6.5 Nm

3.3 No; Forces in AB and CD each 10.31 kN

3.4 1.7 kN; 34.9 kN @ 72° to horizontal

EXERCISES 4

4.1 2 kN, 3 kN, 9 kNm; 50 kN, 34.6 kN, 52 kN

4.2 31.6 kN, 28.6 kN, 16.7 kN

4.3 61.5 kN, 65.6 kN, 36.4 kN

4.4 1207 N, 1344 N, 776 N

Index